Prelude to
Equations

Steven R. Lay
Professor of Mathematics
Lee University

L. Clark Lay
the late Professor of Mathematics Education
California State University, Fullerton

Contributing Artist: Izumi Shimizu

Cover Photo: Ruins of the Abbey in Jedburgh, Scotland
by Tom Talbert

Copyright © 2010 Steven R. Lay

Published by:
The Mathematics Division of
Cross Product Publications
Cleveland, TN

ISBN-13: 978-0-9826837-0-5

Printed in the United States of America.

10 9 8 7 6 5 4 3 2 1
14 13 10 11 10

Contents

Preface

For Students (and Parents of Students)

The transition from arithmetic to algebra is widely recognized as a major stumbling block for many students. In our more than 50 years of combined experience teaching algebra, we have found the single most common reason that students struggle in algebra is that mathematics no longer makes sense to them. Somewhere along the way they stopped trying to understand math and they began to rely on memorizing rules. This approach may be effective in the short run, but after a while the rules get mixed up and the wrong rule is applied in the wrong place at the wrong time. While the deficiencies of this approach are seen most clearly in algebra, the causes of this failure can often be traced back to the earlier grades.

This text is based on the very successful *Prelude* approach to preparing for algebra that emphasizes understanding mathematics, not memorizing rules. The first book in the series, *Prelude to Algebra*, was developed for eighth grade pre-algebra classes. It explains the basic arithmetic operations in a new way that develops the thought patterns students will need in algebra. Some of the topics in *Prelude to Algebra* are typically introduced much earlier in the curriculum. Rather than just reviewing them in the eighth grade from a new perspective, it is helpful to introduce the *Prelude* approach in a natural way when the topics are first encountered. This has led to the writing of several "prelude" books for the lower grades: *Prelude to Signed Numbers* (grade 7), *Prelude to Percents* (grade 6), *Prelude to Equations* (grade 5), and *Prelude to Fractions* (grade 4).

The *Prelude* approach teaches the student to think about arithmetic in a different way that emphasizes the way numbers change and anticipates the operations used in algebra.

- Conceptually, it makes better sense because it represents a dynamic model that corresponds more closely to the way people think about numbers.[1]
- Practically, it illustrates the way that arithmetic and algebra are applied in the real world.[2]
- Mathematically, it is more accurate than the traditional approach.[3]
- Historically, it has been remarkably helpful in decreasing students' fear of mathematics and increasing their success in algebra.[4]

While *Prelude to Equations* teaches new patterns of thinking about arithmetic, it does not significantly change what is written down on paper. If students change schools and do not continue with the *Prelude* curriculum in subsequent years, they are not penalized for writing strange things on their papers or "doing it differently." The difference is in their way of thinking and in their success in subsequent classes. For students who continue with the *Prelude* series, the present book introduces the basic concepts and prepares them to gain a deeper understanding as they revisit these topics in later years.

For the first six chapters (more than half of the book) we work entirely with positive whole numbers and zero—no fractions and no negative numbers. This enables the students with a weaker background to regain their self-confidence in doing math. In this simple setting we will ask significant questions, many of which will be new to all the students—even the most advanced.

By thinking carefully about the basic properties of arithmetic, we develop patterns that anticipate the procedures of algebra. For example, a clear understanding of how the arithmetic equation $5 - 2 = 3$ relates to the equation $2 = 5 - 3$ enables one to solve the algebraic equation $a - x = b$ for x in one step. Similarly, if one can change the equation $5(3 + 4) = 35$ into the equation $4 = \frac{35}{5} - 3$, then solving the equation $a(b + x) = c$ for x is not so intimidating.

Learning mathematics is not a spectator sport. Each lesson introduces a concept, carefully explains its usage, and gives several examples. But the real learning takes place when the students do the exercises. This is where they see for themselves how math works and begin to train their minds to think in logical ways. In the "Develop Your Skill" section of the Exercises, the students are given ample opportunity to gain confidence in the new material from the lesson. Then in the "Maintain Your Skill" section, they practice using key concepts from earlier lessons. This spiral approach to learning has been found to be very effective in increasing understanding and retention of important skills.

Steven R. Lay

slay@leeuniversity.edu

A Special Note to Parents:

If you are used to helping your child with his/her homework, that's great. But be careful. Many of the problems may look familiar to you, but we are teaching the students to think about them in a special way. Before you jump in with the "usual" explanation, you are strongly encouraged to read the lesson to see what we are doing. If you do this on a regular basis, you may even learn a few new things yourself!

---

[1] Before children encounter -1 as a mysterious negative number, they are already comfortable with the concept of decreasing by 1 if they have ever ridden on an elevator or had to share a toy with a friend.

[2] Many applications of algebra involve quantities that are changing. By emphasizing the changes involved in arithmetic computations, we develop a pattern that is easily followed when solving algebraic equations.

[3] For example, traditional texts are consistently inaccurate or unintuitive when defining exponents and frequently misleading when talking about canceling.

[4] The *Prelude* approach has been used for more than 20 years in remedial classes in college. Long-term studies have found that 94% of the students who completed a course using this approach were able to be successful in algebra and other courses that used algebraic skills. Hearing those students make comments such as "Why didn't they explain it this way in middle school?" has prompted the writing of this text.

About the authors:

Dr. L. Clark Lay began his teaching career in a one-room schoolhouse in southern Iowa during the Great Depression. To better identify with the children in his classes, he often walked to school barefoot. During the summers he traveled to California and earned a master's degree in mathematics from the University of Southern California. This lead to his moving to California and teaching high school in El Centro and then junior college in Pasadena. It was at Pasadena City College that he first became interested in helping World War II veterans re-enter an academic world and become successful in algebra, calculus and higher math. These returning veterans were highly motivated, but their math skills were largely undeveloped. This caused a high failure rate in remedial courses in beginning algebra.

The math department surveyed existing arithmetic textbooks, but could find none whose goals specifically included preparing students for algebra. Convinced that such an approach was needed, Lay began a study of the kinds of mistakes that students commonly make in algebra. Then he looked for ways to develop thought patterns in arithmetic that would reduce those algebraic errors. His research led to a doctor's degree in mathematics education from the University of California at Los Angeles and the course he developed at Pasadena City College greatly increased the students' chances for success in mathematics.

In 1960 he became the chairman of the department of mathematics education at California State University at Fullerton, where he trained elementary and middle school math teachers until his retirement in 1975. Over the years, he gained additional insights, made further refinements in the basic approach, and authored several books.

Dr. Lay has two sons, both mathematicians. His older son David is a Professor of Mathematics at the University of Maryland. His younger son Steven has a similar position at Lee University in Tennessee. Both are accomplished authors like their father.

Other books by L. Clark Lay include *Arithmetic, an Introduction to Mathematics* (Macmillan, 1960), *The Study of Arithmetic* (Macmillan, 1966), *From Arithmetic to Algebra* (Macmillan, 1970), *Principles of Elementary Mathematics* (1973), and *Principles of Algebra* (1980, 1985, 1988, 1990).

Dr. Steven R. Lay began teaching at Aurora University (Illinois) in 1971, after earning a Ph.D. in mathematics from the University of California at Los Angeles. He, too, wrestled with the question of how best to teach remedial algebra to under-prepared college students. When his father retired in 1975 and moved to Illinois, they began a joint project of reworking and rewriting the material that had proved so successful in California. This resulted in a new text, *Principles of Algebra*, which they continued to revise and improve for a number of years.

Steven Lay's career in mathematics was interrupted for 8 years while serving as a Christian missionary in Japan. Upon his return to the States in 1998, he joined the mathematics faculty at Lee University in Cleveland, Tennessee. Once again the need for a better remedial algebra curriculum prompted him to adapt the text he and his father had developed in Illinois. This new version concentrated more on the transition between arithmetic and algebra and less on algebra itself.

Over the years as college students have used the material in the Lays' books, they have often made comments such as: "For the first time in my life, math makes sense to me." "Why didn't they explain it this way in middle school?" or "You ought to write a middle school math book so that students can learn it right the first time." This text is a response to those suggestions.

Steven and his wife Ann have two children. Their daughter BethAnn taught 8th grade math for several years and now teaches part-time at Lee University. Their son Tim is pursuing a Ph.D. in history.

Other books by Steven R. Lay include *Convex Sets and their Applications* (John Wiley and Sons, 1982; Dover Publications, 2007), *Analysis with an Introduction to Proof* (Prentice Hall, 1986, 2005), *Japanese Language and Culture* (2003, 2010), *Principles of Algebra* (1980, 2009), *Prelude to Algebra* (2007), *Prelude to Signed Numbers* (2009), *Prelude to Percents* (2009), and *Prelude to Fractions* (2010).

Prelude to
Equations

Chapter 1 — Sums and Differences

1.1 – SYMBOLS AND SUMS

OBJECTIVE

1. Explain the meaning of a sum.

The most common numbers, often called the **natural numbers**, are the whole numbers, starting with 1:

$$1, 2, 3, 4, 5, \ldots .$$

(The three dots mean we continue in the same way.)

We use the natural numbers when we count objects. For example:

> Bill wants to know how many rocks he has in his rock collection. He uses the natural numbers to count them. 1, 2, 3, 4, 5, 6. Bill has 6 rocks in his collection.

Sometimes it is useful to use the number zero. It gives us a place to start.

> How many rocks did Bill have before he started collecting rocks? Before he started collecting, he had zero rocks.

Yeah! No fractions!

When we include zero with the natural numbers, we get what are called the **counting numbers**:

$$0, 1, 2, 3, 4, \ldots .$$

For most of this book, we will use only the counting numbers, and no fractions. Later when we introduce fractions (in Chapter 7), the rules for computing with them will be much easier.

Sometimes we want to talk about a number without saying exactly what number it is. In this case we may represent the number by a letter. For example, we may write

Let x be a counting number.

In this case the letter x is used as a **variable**, or placeholder. It holds in reserve a place in the sentence.

Example 1   Using a Variable

Q: If x is a counting number, what numbers could take the place of x?

A: x could be 0, 1, 2, 3, or any other counting number.

Q: Could x be $\dfrac{1}{2}$?

A: No, because $\dfrac{1}{2}$ is not a counting number.

If we choose two counting numbers, say 4 and 7, we can add these two numbers together to get their **sum** $4 + 7$. In this case, we call addition the **operation** we performed on the two numbers. The operation tells us how the two numbers have been combined: we added them together.

Definition of a Sum

If x and y are counting numbers, then addition assigns to x and y a unique third counting number $x + y$, called the **sum** of x and y.

Example 2   Find a Sum

Find the sum of each pair of counting numbers.

  (a) 5 and 3   (b) 4 and t (c) m and n

Using the definition above, we obtain the following:

  (a) $5 + 3$ (b) $4 + t$ (c) $m + n$

While the definition states that $5 + 3$ is the sum of 5 and 3, it is also true that 8 is the sum of 5 and 3. The numerals $5 + 3$ and 8 both represent the number eight. In fact, there are many ways to show the number eight. Here are just a few:

$$8, \; 5 + 3, \; 6 + 2, \; 4 + 4, \; 7 + 1$$

Of the many possible ways to write the number eight, the simplest way is to write 8. We call this form the **basic numeral** for the number eight.

We can state that $5 + 3$ and 8 are names for the same number by writing the equation

$$5 + 3 = 8.$$

The equal sign means that $5 + 3$ and 8 have the same value. When we replace $5 + 3$ by 8 we are doing an addition **computation**. Replacing $5 + 3$ with 8 is a change in form only and *not* a change in the number represented.

Key Point

> A **computation** is a change of form where an operation is performed and the result is the basic numeral.

> ### Example 3   Compute a Sum
>
> Compute the basic numeral for each sum.
>
>   (a) $3 + 6$ (b) $8 + 5$ (c) $x + 4$
>
> (a)   The basic numeral for $3 + 6$ is 9, because $3 + 6 = 9$.
> (b)   The basic numeral for $8 + 5$ is 13, because $8 + 5 = 13$.
> (c)   Since the value of x is not known, it is not possible to compute.

In Example 3(c) it was not possible to compute the basic numeral for the sum $x + 4$. This is because x is just a symbol for a blank space. Unless x is replaced by a particular number, no computation is possible. In this case the word "sum" is used as the *name of a form*. That is, "sum" describes the appearance of $x + 4$, even though no computation is possible. Similarly, we say that $5 + 3$ is a sum because of its form.

EXERCISE 1.1

DEVELOP YOUR SKILL

Indicate whether each statement in Exercises 1 – 4 is True or False.

1. 5 is a natural number.   **2.** 8 is a counting number.

3. 1 is a counting number.   **4.** 0 is a natural number.

5. What, by definition, is the sum of r and s?

6. What, by definition, is the sum of t and t?

7. What, by definition, is the sum of 6 and 2?

8. What, by definition, is the sum of 3 and 7?

Compute the basic numeral for each sum. If no computation is possible, write "not possible."

9. $15 + 4$ **10.** $23 + 43$ **11.** $32 + x$

12. $37 + 22$ **13.** $13 + 40$ **14.** $7 + 18$

15. $23 + 27$ **16.** $y + 13$ **17.** $28 + 19$

18. $15 + x$ **19.** $35 + 44$ **20.** $45 + 32$

1.2 – EVEN NUMBERS

OBJECTIVE

1. Show why a given number is even.

When a number is added to itself, the result is called an even number. For example, $5 + 5 = 10$, so 10 is an even number. This shows that an even number can always be divided exactly by 2. When 10 is divided by 2, we get 5. And 5 is what must be added to itself to get 10. The even numbers are easy to identify since they always end in one of the digits 0, 2, 4, 6, or 8. But to show why a number is even, we want to write the addition equation that makes it even. Here is the general rule:

Definition of an Even Number

A counting number x is an **even** number if there is a counting number n such that

$$n + n = x.$$

When a variable is used in more than one place in an equation, the same number must be substituted for the variable wherever it occurs. For example, in the equation $n + n = x$, if one of the n's is replaced by a number, then the other n must be replaced by the same number. Since x is a different letter from n, x and n can be replaced by different numbers. But the same number can replace both x and n, if that is desired.

If a counting number is not even, then it is said to be **odd**. The odd numbers end in one of the digits 1, 3, 5, 7, or 9. If x is an odd number, then it cannot be divided exactly by 2 and so there is no counting number n such that $n + n = x$.

> **Example 1   Show a Number is Even**
>
> *Show that* 0, 2, *and* 4 *are each even numbers by writing the equations required by the definition of an even number.*
>
> 0 is even since $0 + 0 = 0$.
>
> 2 is even since $1 + 1 = 2$.
>
> 4 is even since $2 + 2 = 4$.

We must be careful when we use a definition to show something is true. We have to use the definition exactly as it is written. In Example 1 we said that 4 is even since $2 + 2 = 4$. It is also true that $4 = 2 + 2$, but that is not exactly the way the equation appears in the definition. The equation in the definition is $n + n = x$, where the adding is on the left side of the equation. So the correct way to show that 4 is even is to write $2 + 2 = 4$, with the adding on the left side.

EXERCISE 1.2

DEVELOP YOUR SKILL

Show that each statement is true by writing the equation required by its definition.

1. 12 is an even number.
2. 6 is an even number.
3. 0 is an even number.
4. 22 is an even number.
5. 28 is an even number.
6. 48 is an even number.
7. 30 is an even number.
8. 26 is an even number.

MAINTAIN YOUR SKILL

9. What, by definition, is the sum of 6 and 7? [1.1]
10. What, by definition, is the sum of m and 8? [1.1]

Compute the basic numeral for each sum. If no computation is possible, write "not possible." [1.1]

11. 30 + 15   **12.** 35 + 23

13. 16 + x **14.** 26 + 14

15. 25 + 28   **16.** n + 29

17. 27 + 16   **18.** 15 + 22

19. 31 + 43   **20.** 23 + y

21. 42 + a **22.** 32 + 26

23. 38 + 23   **24.** 22 + 65

1.3 – LESS THAN INEQUALITIES

OBJECTIVE

1. Explain less than inequalities.

   When two counting numbers are not equal to each other, then one number will be smaller than the other number. For example, we know that 3 is less than (smaller than) 5. If we place 3 blocks on one side of a balance and 5 on the other side, we see that the 3 blocks weight less than the 5 blocks. We say that 3 is less than 5 by writing 3 < 5.

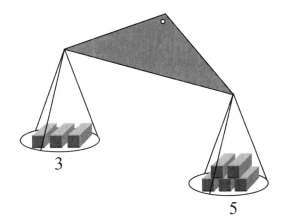

3

5

How can we show that $3 < 5$? What is it about 3 that makes it less than 5? The number 3 is less than 5 because we have to add something to 3 to make it equal to 5. What do we add? Of course, we add 2 to 3 to make it equal to 5.

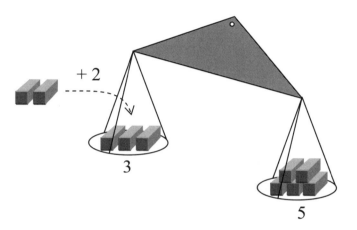

So we can say $3 < 5$ because $3 + 2 = 5$. This is the idea behind the following definition of "less than."

Definition of Less Than	Let x and y be counting numbers. We say that x is less than y and we write $x < y$ if there is a natural number n such that $$x + n = y.$$

In the definition of "less than," the number n is required to be a natural number, not a counting number. This means that the number we add to x is not zero.

Example 1   Explain a "Less Than" Inequality

Show that 2 < 9 by writing the equation required by the definition.

Since 2 is smaller than 9, we can add a number to the 2 to make it equal to 9. What do we need to add? Of course, we add 7. Keep the two numbers, 2 and 9, in the same order, and add the 7 to the smaller side. The desired equation is

$$2 + 7 = 9.$$

This equation proves that 2 < 9, because it is the equation that is required by the definition.

Notice the order of the numbers on the left side of the equation in the solution to Example 1. We have to be very precise in using a definition. While it is true that $7 + 2 = 9$, this equation is not the equation required by the definition of "less than." The definition adds the natural number n (in this case 7) to the right side of 2. So we must write

$$2 + 7 = 9$$

and not $7 + 2 = 9$. We shall see later that $x + n$ and $n + x$ always represent the same number, but in terms of their form, they are different.

An equation, such as $2 + 7 = 9$, is also called an **equality**. The relation 2 < 9 is an example of an **inequality**. The relation of "less than or equal to" is also called an inequality. The notation $x \leq y$ means that $x < y$ or $x = y$.

Study Tip:

Note in (a) that the 6 stays on the left side and the 8 stays on the right side. And the 2 is added to the right of the 6 to make it into an equation.

Example 2   Explain Inequalities

Show that each inequality is true by writing the equation required by its definition.

  (a) $6 < 8$ (b) $5 \leq 7$ (c) $3 \leq 3$

(a) We must add 2 to 6 to make it equal to 8: $6 + 2 = 8$.

(b) 5 is not equal to 7, but it is less than 7. To show it is less than 7 we write

$$5 + 2 = 7.$$

(c) 3 is not less than 3, but they are equal. We show this by writing

$$3 = 3.$$

EXERCISE 1.3

DEVELOP YOUR SKILL

Show that each inequality is true by writing the equation required by its definition.

1. $4 < 9$ 2. $3 \leq 5$

3. $5 < 7$ 4. $11 < 15$

5. $0 < 5$ 6. $2 < 3$

7. $5 \leq 9$ 8. $4 \leq 7$

9. $6 \leq 6$ 10. $13 < 45$

11. $2 < 17$ 12. $8 \leq 8$

MAINTAIN YOUR SKILL

Compute the basic numeral for each sum. If no computation is possible, write "not possible." [1.1]

13. $17 + c$ 14. $21 + 17$

15. $16 + 13$ 16. $26 + 14$

17. $12 + 11$ **18.** $31 + x$

19. $38 + 14$ **20.** $n + 5$

21. Joe had $235. Jim paid Joe $58 that he owed him. Then how much money did Joe have? [1.1]

Show that each statement is true by writing the equation required by its definition. [1.2]

22. 22 is an even number.   **23.** 34 is an even number.

24. 46 is an even number.   **25.** 50 is an even number.

1.4 – GREATER THAN INEQUALITIES

OBJECTIVE

1. Explain greater than inequalities.

The inequality $3 < 9$ tells us that 3 is less than 9. But if 3 is less than 9, then 9 is greater than 3. We write this as $9 > 3$. In this case, the smaller number is on the right side, so to make them equal, we have to add something to the right-hand side.

Definition of

Greater Than

Let x and y be counting numbers. We say that x is greater than y and we write $x > y$ if there is a natural number n such that

$$x = y + n.$$

To say that x is greater than or equal to y we write $x \geq y$. This means that $x > y$ or $x = y$.

Study Tip:

Note that the answer in (a) is
$5 = 2 + 3$
and not
$5 = 3 + 2$
or $2 + 3 = 5$.
The form of the definition must be followed exactly.

Example 1   Explain Inequalities

Show that each inequality is true by writing the equation required by its definition.

    (a) $5 > 2$ (b) $10 > 2$ (c) $12 \geq 3$ (d) $4 \geq 4$

(a) Since 5 is larger than 2, something must be added to 2, namely 3, to make it equal to 5. Keep the two numbers, 5 and 2, in the same order and add the 3 to the smaller side. The desired equation is

$$5 = 2 + 3.$$

(b) To have 10, we need to add 8 to the 2: $10 = 2 + 8$.

(c) 12 is not equal to 3, but it is greater than 3. We show this by writing

$$12 = 3 + 9.$$

(d) 4 is not greater than 4, but they are equal. To show they are equal we write

$$4 = 4.$$

EXERCISE 1.4

DEVELOP YOUR SKILL

Show that each statement is true by writing the equation required by its definition.

  1. $6 > 2$ **2.** $10 > 5$

  3. $7 > 1$ **4.** $11 > 8$

  5. $12 \geq 3$ **6.** $5 > 4$

  7. $18 > 14$ **8.** $4 \geq 0$

  9. $25 > 21$ **10.** $17 > 5$

11. $23 \geq 23$ **12.** $34 > 29$

MAINTAIN YOUR SKILL

13. What, by definition, is the sum of 12 and 12? [1.1]

14. What, by definition, is the sum of 3 and c? [1.1]

Compute the basic numeral for each sum. If no computation is possible, write "not possible." [1.1]

15. $18 + 27$ **16.** $36 + 18$

17. $26 + x$ **18.** $22 + 35$

Show that each statement is true by writing the equation required by its definition. [1.2, 1.3]

19. 24 is an even number.   **20.** 40 is an even number.

21. $2 < 27$ **22.** $4 \leq 18$

23. $29 \leq 29$ **24.** $35 < 38$

1.5 – DIFFERENCES

OBJECTIVE

1. Identify the relationship between sums and differences.

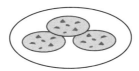 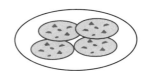

Suppose we have two plates with chocolate chip cookies on them. One plate has 3 cookies and one plate has four cookies. If we eat all the cookies on both plates, we will have eaten a total of seven cookies. We write this as

$$3 + 4 = 7.$$

But suppose we only eat the cookies on the plate of three. How many cookies are left? We can see that there are four left. We write this as

$$7 - 3 = 4.$$

These two equations show how the two parts 3 and 4 are related to the whole 7. If we take the two parts and add them together, we get the whole. $3 + 4 = 7$. If we start with the whole amount and subtract one part, we get the other part. $7 - 3 = 4$.

Whenever we have two counting numbers, we can always add them together and get another counting number. For example, 3 and 4 are counting numbers, and their sum $3 + 4$ is the counting number whose basic numeral is 7.

Subtraction is different. If we have the counting numbers 3 and 7, we can subtract 3 from 7 and get the counting number $7 - 3$ whose basic numeral is 4. Where did the 4 come from? It is the counting number that must be added to 3 to get 7. In other words,

$$7 - 3 = 4 \quad \text{because} \quad 3 + 4 = 7.$$

But if we try to subtract 12 from 7, we encounter a problem. If we have 7 cookies, can we eat 12 of them? No, we can't, because 12 is greater than 7. To follow the pattern above, if we could subtract 12 from 7, then we would have to be able to add something on to 12 and get 7.

$$12 + \boxed{?} = 7$$

But this is not possible, because 12 is already larger than 7. There is no counting number that we can add to 12 to make it equal to 7. That means we cannot subtract 12 from 7 and get a counting number.

We say that subtraction is a **partial operation** on the counting numbers. Sometimes we can do it and sometimes we cannot. In order to subtract a number y from a number x, the number x must be at least as large as y. That is, in order for $x - y$ to be defined, we must have $x \geq y$. The following definition describes subtraction by showing its relationship to addition.

Definition of Subtraction	Let x and y be counting numbers. If $x \geq y$, then y can be subtracted from x, and we write $$x - y = z,$$ where z is the counting number such that $y + z = x$. The expression $x - y$ is called the **difference** of x and y.

Study Tip:

Notice that y is to the left of z in both the subtraction equation and the addition equation.

The definition of subtraction shows the relationship between subtraction and addition. We see that

$$x - y = z \quad \text{corresponds to} \quad y + z = x.$$

For example,

$$10 - 6 = 4 \quad \text{corresponds to} \quad 6 + 4 = 10.$$

Likewise,

$$8 - 2 = 6 \quad \text{corresponds to} \quad 2 + 6 = 8$$

and

$$14 - 5 = 9 \quad \text{corresponds to} \quad 5 + 9 = 14.$$

We could also begin with an addition equation and pair it with a subtraction equation. For example,

$$4 + 8 = 12 \quad \text{corresponds to} \quad 12 - 4 = 8.$$

Notice that the order of the numbers is important here. If

$$y + z = x \quad \text{is represented by} \quad 4 + 8 = 12,$$

then $y = 4$, $z = 8$, and $x = 12$. This means the subtraction equation $x - y = z$ becomes $12 - 4 = 8$. While it is also true that $12 - 8 = 4$, this would not be the subtraction equation that corresponds with $4 + 8 = 12$, according to the definition.

Example 1   Relate Subtraction to Addition

Use the definition of subtraction to change each subtraction equation into the corresponding addition equation.

(a) $76 - 12 = 64$ (b) $a - b = c$

(a) The 12 and the 64 must be added together to get their sum 76. Keep them in the same order and put them on the left side:

$$12 + 64 = 76.$$

(b) This time we don't have numerical values, so we just have to follow the form of the definition. If $a - b = c$, then b and c are added together to get a. Keep the b and the c in the same order as the original equation: b is to the left of c. We have

$$b + c = a.$$

Example 2   Relate Addition to Subtraction

Use the definition of subtraction to change each addition equation into the corresponding subtraction equation.

(a) $25 + 38 = 63$ (b) $m + n = p$

(a) 63 is the sum of 25 and 38, so when we subtract 25 from 63 we get 38. Note that 25 is to the left of 38 in both equations:

$$63 - 25 = 38.$$

(b) Since $m + n = p$, if we subtract m from p we will get n:

$$p - m = n.$$

Note that m is to the left of n in both equations.

Study Tip:

The expression
$x - y$
is a difference,
because of its
form. Just like
$x + y$
is a sum.

In Lesson 1.1 we saw that the word "sum" can have two meanings. It may refer to the answer to an addition problem, or it may refer to the form of an expression. The same is true for the word "difference." In the equation $x - y = z$ we may refer to z as the difference that results from the subtraction computation x minus y. But we may also refer to $x - y$ as a difference because of its form.

Example 3   Use the Definition of Subtraction

Billy bought a basketball for a cost (C) of $5.00. When he paid for it, he also had to pay $0.45 in sales tax (T). The final price (P) for the basketball was $5.45. We can show this by the equation C + T = P. Use the definition of subtraction to write an equation that shows how to find the tax (T), when you know the cost (C) and the price (P).

The addition equation $C + T = P$ corresponds to the subtraction equation $P - C = T$. This shows how to find the tax T when the purchase price P and the cost C are given: subtract C from P.

EXERCISE 1.5

DEVELOP YOUR SKILL

Compute the basic numeral.

1. $12 - 7$ 2. $15 - 9$ 3. $18 - 11$
4. $21 - 17$ 5. $23 - 15$

Use the definition of subtraction to change each subtraction equation to the corresponding addition equation, and each addition equation to the corresponding subtraction equation.

6. $8 - 2 = 6$ 7. $5 + 9 = 14$

8. $11 + 6 = 17$

9. $18 - 15 = 3$

10. $21 - 7 = 14$

11. $18 + 12 = 30$

12. $5 + m = 14$

13. $n - 12 = 6$

14. $b - c = 23$

15. $41 - x = m$

16. $a + h = y$

17. $m - x = k$

18. If the interest i is added to the principle P, this sum is the amount A to pay at the close of a loan. We may represent this by $P + i = A$. Use the definition of subtraction to write an equation that shows how to find the interest i when you know the amount A and the principle P.

MAINTAIN YOUR SKILL

Show that each statement is true by writing the equation required by its definition.

[1.2, 1.3, and 1.4]

19. 16 is an even number.

20. 30 is an even number.

21. $6 < 17$

22. $7 \leq 15$

23. $18 > 3$

24. $13 < 16$

1.6 – ORDER OF OPERATIONS

OBJECTIVE

1. Use the order of operations.

Some math expressions have more than one operation:

$$11 - 6 + 3$$

One student might compute this by thinking

"11 minus 6 equals 5, and 5 plus 3 equals 8."

Another student might think,

"6 plus 3 equals 9, and 11 minus 9 equals 2."

Who would be correct? The first student is right because we always work from left to right, the same as the way we read. If we want to change this left to right order, we use parentheses. When you see parentheses in a math sentence, always do what is inside the parentheses first.

Order of Operations for Sums and Differences

When computing with sums and differences, work from left to right, unless told to do otherwise by parentheses. When parentheses are used, do the operation inside the parentheses first.

For example, $10 - 7 + 2$ is computed as $10 - 7 = 3$ and then as $3 + 2 = 5$. This would be the same as if it were written as

$$
\begin{aligned}
10 - 7 + 2 &= (10 - 7) + 2 \\
&= \quad 3 \ + \ 2 \\
&= \quad\quad 5
\end{aligned}
$$

In the expression $10 - (7 + 2)$, the $7 + 2$ is computed first because it is inside the parentheses:

$$
\begin{aligned}
10 - (7 + 2) &= 10 - 9 \\
&= 1
\end{aligned}
$$

Notice that $10 - 7 + 2 \neq 10 - (7 + 2)$.

Study Tip:

The symbol " $\neq$ " means "is not equal to."

Example 1   **Use the Order of Operations**

Compute the basic numeral.

  (a) $12 - 6 + 2$ (b) $12 - (6 + 2)$

  (c) $16 - 5 - 1$ (d) $16 - (5 - 1)$

(a) $12 - 6 + 2 = (12 - 6) + 2 = 6 + 2 = 8$

(b) $12 - (6 + 2) = 12 - 8 = 4$

(c) $16 - 5 - 1 = (16 - 5) - 1 = 11 - 1 = 10$

(d) $16 - (5 - 1) = 16 - 4 = 12$

Some differences cannot be computed when using the counting numbers. For example, if you only have 5 pencils, can you give me 7? Of course not! $5 - 7$ cannot be computed when using just the counting numbers. We say that $5 - 7$ is **undefined**. Remember that for a subtraction sentence to be defined, the first number must be great than or equal to the second number.

Remember:

The difference
$x - y$
is defined only
when $x \geq y$.

Example 2   Compute the Basic Numeral

Compute the basic numeral for each expression that is defined as a counting number. Otherwise, write "not defined."

(a) $8 - 3 + 6$ (b) $8 - (3 + 6)$

(c) $4 - 6 + 9$ (d) $5 - 2 - 6$

(a) We work from left to right:
$$8 - 3 + 6 = (8 - 3) + 6 = 5 + 6 = 11$$

(b) We begin inside the parentheses and add 3 and 6:
$$8 - (3 + 6) = 8 - 9.$$
But now we are asked to subtract 9 from 8, and this is not defined. So the original expression is not defined either.

(c) Since we work from left to right, we must begin with $4 - 6$. This first step is not defined, so the whole expression is not defined.

(d) We begin by subtracting 2 from 5:
$$5 - 2 - 6 = (5 - 2) - 6 = 3 - 6.$$
But now we are asked to subtract 6 from 3, and this is not defined. So the original expression is not defined either.

EXERCISE 1.6

DEVELOP YOUR SKILL

Compute the basic numeral for each expression that is defined as a counting number. Otherwise, write "not defined."

1. $12 - 8 + 2$

2. $11 - (3 + 4)$

3. $10 - (2 + 5)$

4. $15 - 7 - 4$

5. $6 - 8 + 5$

6. $9 - 7 - 2$

7. $7 + 4 - 8$

8. $(5 - 8) + 4$

9. $6 - 9 + 4$

10. $7 + 5 - 6$

11. $(9 - 7) + 5$

12. $13 - 9 - 5$

MAINTAIN YOUR SKILL

Show that each statement is true by writing the equation required by its definition.
[1.2. 1.3, and 1.4]

13. 26 is an even number.

14. $15 < 18$

15. $17 \geq 13$

16. $26 > 19$

17. $25 \leq 25$

18. $30 \geq 24$

19. There were 55 students performing at the half-time of the football game. Sixteen of them were in the drill team and the rest were in the band. How many students were in the band? [1.5]

Use the definition of subtraction to change each subtraction equation to the corresponding addition equation, and each addition equation to the corresponding subtraction equation. [1.5]

20. $13 - 4 = 9$

21. $14 + 21 = 35$

22. $x - 5 = 18$

23. $20 + m = 32$

24. $6 + y = n$

25. $h - k = 20$

1.7 – PROPERTIES OF SUMS

OBJECTIVES

1. Use the commutative and associative principles for addition.
2. Recognize the identity property for addition.

When we write a math expression like $x + y$ or $x - y$, the x and y are called **terms**. What happens if we interchange the order of the terms? Will we get an equivalent expression? We know that

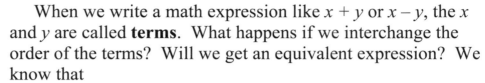

$$3 + 7 = 10 \quad \text{and} \quad 7 + 3 = 10$$

This illustrates the **commutative principle for addition**. When we add numbers together, the order does not matter. We can state this as a general rule:

Key Concept

> If x and y are counting numbers, then $x + y = y + x$.

If we are subtracting and we change the order of the terms, will we get the same answer? For example, we know that $7 - 3$ is equal to 4. But $3 - 7$ is not even defined. So, subtraction is *not* commutative.

Example 1   Use the Commutative Principle

Use the commutative principle to replace each expression by one or more equivalent expressions. Do not compute.

(a) $4 + 17$ (b) $12 + (8 - 3)$

(a) $17 + 4$

(b) This time we treat $(8 - 3)$ as a term and interchange it with 12. The only possibility is $(8 - 3) + 12$.

Study Tip:

When we "commute" to work, we travel back and forth. We change places from our home to our job. In using the commutative principle for sums, the two terms on either side of the sum change places; they commute.

Example 2   Use the Commutative Principle

Use the commutative principle to replace each expression by one or more equivalent expressions, if possible.

(a) $n - (x + 3)$ (b) $y - 7$ (c) $r + (s + t)$

(a) We can only commute the x and the 3:

$$n - (3 + x)$$

(b) Since subtraction is not commutative, it is not possible to make any changes using the commutative principle.

(c) This time there are several possibilities:

We may commute the s and t within the last term:

$$r + (t + s).$$

We may commute the r and the term $(s + t)$:

$$(s + t) + r.$$

We may make both changes: $(t + s) + r.$

There is another way to change the expression $r + (s + t)$ in Example 2(c). We may leave the r, s, and t written in the same order, but group the r and s together with parentheses, instead of the s and t. This changes the order of the addition operations. Instead of adding s and t and then adding r, we can add r and s, and then add t. This new expression, $(r + s) + t$, cannot be obtained from $r + (s + t)$ by using the commutative principle. But is it equivalent to the original? Yes, it is. For example,

$$5 + (3 + 7) = 5 + 10 = 15, \quad \text{and}$$

$$(5 + 3) + 7 = 8 + 7 = 15.$$

This is called the **associative principle for addition**: when adding more than two terms together, it does not matter how you group those terms.

Key Concept

If x, y, and z are counting numbers, then

$$x + (y + z) = (x + y) + z.$$

Study Tip:

In using the associative principle, the terms stay in the same order, they are just grouped (associated) differently.

Example 3   **Identify Commutative and Associative Principles**

State whether the commutative principle or the associative principle is used in making each change of form.

(a) $(5 + 6) + 9 \rightarrow 5 + (6 + 9)$

(b) $(5 + 6) + 9 \rightarrow (6 + 5) + 9$

(c) $(5 + 6) + 9 \rightarrow 9 + (5 + 6)$

(a) The numbers stay in the same order, but they are grouped (associated) differently. The associative principle is used.

(b) The terms 5 and 6 have been commuted.

(c) The terms $(5 + 6)$ and 9 have been commuted.

   The number zero has special properties for addition and subtraction. When zero is added to any number, the sum is always the same as the original number. Similarly, when any number is added to zero, the sum is again that same number. No number other than zero satisfies these conditions. This is known as the **identity property for addition**:

Identity Property for Addition	For any counting number x, $$x + 0 = x \quad \text{and} \quad 0 + x = x.$$

If we use the definition of subtraction to write the corresponding subtraction equations, we have

$$x - x = 0 \quad \text{and} \quad x - 0 = x.$$

EXERCISE 1.7

DEVELOP YOUR SKILL

Write all other possible equivalent forms that result from using only the commutative principle for addition. If there are none, write "none."

1. $a + (7 + b)$
2. $m - (3 + p)$
3. $m + (n - 5)$
4. $(r + 4) - t$
5. $(x - y) + 6$
6. $r - (s - 2)$

State whether the commutative or associative principle is used in making each change of form: (a) to (b), (b) to (c), etc.

7. (a) $(2 + 3) + 6$
    (b) $(3 + 2) + 6$
    (c) $3 + (2 + 6)$
    (d) $(2 + 6) + 3$
    (e) $(6 + 2) + 3$
    (f) $6 + (2 + 3)$

8. (a) $z + (y + x)$
    (b) $(z + y) + x$
    (c) $x + (z + y)$
    (d) $x + (y + z)$
    (e) $(y + z) + x$
    (f) $y + (z + x)$

MAINTAIN YOUR SKILL

Use the definition of subtraction to change each subtraction equation to the corresponding addition equation, and each addition equation to the corresponding subtraction equation. [1.5]

9. $15 - 2 = 13$
10. $7 + 3 = 10$

11. $21 - n = 5$ **12.** $x + 8 = 22$

13. $9 + m = x$ **14.** $k - 12 = r$

Compute the basic numeral for each expression that is defined as a counting number. Otherwise, write "not defined." [1.6]

15. $4 + 9 - 7$ **16.** $(6 + 10) - 9$

17. $3 - 5 + 6$ **18.** $9 - 7 + 8$

19. $17 - (5 - 2)$ **20.** $10 - 8 - 5$

21. $5 - 7 - 3$ **22.** $6 + 7 - 2$

23. $5 + 1 - 7$ **24.** $16 - 10 - 8$

1.8 – INVARIANT PRINCIPLE FOR SUMS

OBJECTIVE

1. Use the invariant principle for sums.

Suppose we have two bags of marbles. The first bag contains 7 marbles and the second bag contains 3 marbles, for a total of 10 marbles in all. If we take 2 marbles out of the bag of 7 and put them in the bag of 3, then we still have 10 marbles in all— now with 5 marbles in the first bag and 5 in the second. (See Figure 1.1.)

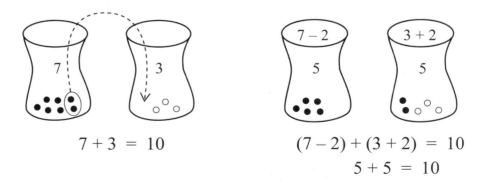

$$7 + 3 \;=\; 10$$

$$(7 - 2) + (3 + 2) \;=\; 10$$
$$5 + 5 \;=\; 10$$

Figure 1.1 Invariant Principle for Sums

This shows that the sum of two numbers is not changed by subtracting something from one number and then adding the same amount to the other number. We call this the **invariant principle for sums**:

Key Concept

Let x, y and z be counting numbers.

If $x \geq z$, then $x + y = (x - z) + (y + z)$.

If $y \geq z$, then $x + y = (x + z) + (y - z)$.

Study Tip:

The word "invariant" means "does not change." In the invariant principle for sums, the sum does not change; it remains invariant.

One use of the invariant principle for sums is to change a given problem into an easier one. For example, it may be difficult to compute $5486 + 198$ in your head without using a notepad or a calculator. But from the invariant principle for sums we have

$$5486 + 198 = (5486 - 2) + (198 + 2)$$
$$= 5484 + 200,$$

and this last sum is easily seen to be 5684. The idea here is to change the "difficult" number 198 into the "easy" number 200. By making the opposite change in the other term, the sum is invariant—it remains the same.

Example 1   Use the Invariant Principle for Sums

Use the invariant principle for sums to simplify each expression. Then compute the basic numeral.

(a) $295 + 637$ (b) $632 + 280$

(a)   295 is only 5 less than 300, so add 5 to 295 and subtract 5 from 637:

$$295 + 637 = (295 + 5) + (637 - 5)$$
$$= 300 + 632 = 932.$$

(b)   280 is 20 less than 300, so add 20 to 280 and subtract 20 from 632:

$$632 + 280 = (632 - 20) + (280 + 20)$$
$$= 612 + 300 = 912.$$

Or, we could subtract 32 from 632 and add 32 to 280:

$$632 + 280 = (632 - 32) + (280 + 32)$$
$$= 600 + 312 = 912.$$

When a variable is used in an equation, it acts like a placeholder. The variable reserves a place for a number. When we replace the variable by a particular number, the resulting equation may or may not be true. For example, in the equation

$$5 + x = 7,$$

if we replace x by 3, then the resulting equation $5 + 3 = 7$ is not true. To **solve an equation** means to find the value of the variable that makes the equation true.

A simple equation like $5 + x = 7$ is easy to solve. We ask ourselves, "What do we add to 5 to get 7?" Of course, we need to add 2. So the solution of $5 + x = 7$, is $x = 2$. That is, when x is replaced by 2 we get the true equation:

$$5 + 2 = 7.$$

Some equations which appear to be quite complicated can actually be solved easily using the invariant principle.

Example 2   **Use the Invariant Principle for Sums**

Solve the equation $452 + 719 = 442 + x$.

Rather than doing a lot of arithmetic, look carefully at the equation to see what is different between the two sides.

$$452 + 719 = 442 + x$$

The term 452 in the sum on the left has decreased by 10 to become 442 in the sum on the right. In order for the sums to remain the same, the other term on the left (719) must be increased by 10. That is,

$$x = 719 + 10 = 729.$$

Example 3   **Use the Invariant Principle for Sums**

Solve the equation $875 + x = 855 + 759$.

The variable x is on the left side, so we find the change from right to left.

$$875 + x = 855 + 759$$

We see that 855 increases by 20 to become 875. To keep the sum the same, 759 must be decreased by 20 to get x. That is, $x = 759 - 20 = 739$.

EXERCISE 1.8

DEVELOP YOUR SKILL

Use the invariant principle to write equivalent sums that are easier to compute.

1. $137 + 98$
2. $295 + 47$
3. $97 + 256$
4. $158 + 396$

Solve each equation by using the invariant principle for sums.

5. $705 + 548 = 707 + x$
6. $x + 283 = 434 + 285$
7. $912 + x = 916 + 567$
8. $235 + 167 = x + 157$

MAINTAIN YOUR SKILL

9.   What, by definition, is the sum of 6 and 4? [1.1]

10. Show that $9 < 16$ by writing the equation required by its definition. [1.3]

Use the definition of subtraction to change each subtraction equation to the corresponding addition equation, and each addition equation to the corresponding subtraction equation. [1.5]

11. $14 + 28 = 42$
12. $51 - 17 = 34$
13. $k - 18 = 23$
14. $15 + c = 32$
15. $15 + y = p$
16. $w + 4 = x$
17. $h - 3 = n$
18. $25 - a = b$

Compute the basic numeral for each expression that is defined as a counting number. Otherwise, write "not defined." [1.6]

19. $3 - 7 + 4$
20. $9 - 6 - 5$
21. $8 - 3 + 6$
22. $11 - 7 + 5$
23. $14 - 12 + 5$
24. $12 + 6 - 10$

1.9 – INVARIANT PRINCIPLE FOR DIFFERENCES

OBJECTIVE

1. Use the invariant principle for differences.

Suppose we decide to sort the books in our library. In the history pile we have 9 books and in the science pile we have 6. We see that we have 3 more history books than science books and observe that $9 - 6 = 3$. If we remove 2 books from each pile, then we will have 7 history books and 4 science books. But we will still have 3 more history books than science books, since $7 - 4 = 3$. (See Figure 2.)

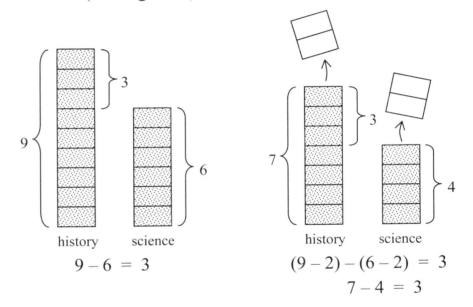

history   science

$9 - 6 = 3$

history   science

$(9 - 2) - (6 - 2) = 3$

$7 - 4 = 3$

Figure 2. Invariant Principle for Differences

The same thing would happen if we added 5 books to the original piles. We would then have 14 history books and 11 science books, but the difference $14 - 11$ would still equal 3.

This shows that the difference between two numbers is not changed if both numbers are either increased or decreased by the same amount. We call this the **invariant principle for differences**:

Key Concept

Let x, y and z be counting numbers.

   If $x \geq y$, then $x - y = (x + z) - (y + z)$.

   If $x \geq y$ and $y \geq z$, then $x - y = (x - z) - (y - z)$.

One use of the invariant principle for differences is to change a given problem into an easier one. For example, it may be difficult to compute $5637 - 298$ in your head without using a notepad or a calculator. But from the invariant principle for differences we have

$$5637 - 298 = (5637 + 2) - (298 + 2)$$
$$= 5639 - 300.$$

We can easily compute this to get 5339. The idea here is to change the "difficult" number 298 into the "easy" number 300. By making the same change in the other term, the difference is invariant—it remains the same.

Example 1   Use the Invariant Principle for Differences

Use the invariant principle for differences to simplify each expression. Then compute the basic numeral.

   (a) $243 - 196$ (b) $453 - 298$

This makes subtracting easier!

(a) To simplify the difference $243 - 196$, we notice that 196 is 4 less than 200. So we add 4 to both terms:
$$543 - 196 = (543 + 4) - (196 + 4)$$
$$= 547 - 200 = 347.$$

(b) To simplify $453 - 298$, we want to add 2 to 298 to make it 300. To keep the difference the same, we add 2 to both terms:
$$453 - 298 = (453 + 2) - (298 + 2)$$
$$= 455 - 300 = 155.$$

Example 2   Use the Invariant Principles

Solve each equation by using an invariant principle.

   (a) $713 - 429 = 813 - y$

   (b) $687 + x = 657 + 479$

Study Tip:

Keeping the difference the same is like subtracting or adding books to each pile. Both piles change in the same way.

(a) Compare the two sides of the equation:

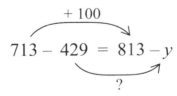

We see that 713 has increased by 100 to become 813. To keep the difference the same, 429 must change into y in the same way. That is, y is 100 more than 429:

$$y = 429 + 100 = 529.$$

Study Tip:

Keeping the sum the same is like moving marbles from one bag to the other. One goes up and the other goes down.

(b) Compare the two sides of the equation:

$$\overset{+\,30}{\overbrace{687 + x \;=\; 657 + 479}_{?}}$$

The variable x is on the left side, so we find the change from right to left. We see that 657 increases by 30 to become 687. To keep the sum the same, 479 must be decreased by 30 to get x. That is, $x = 479 - 30 = 449$.

Example 3   Use an Invariant Principle

Last year Raul's height was 146 cm and Hector's height was 138 cm. If they grew the same amount and Raul's height is now 151 cm, how tall is Hector?

(a) Let x represent Hector's height in cm. Write an equation with a difference on each side that describes this problem.

(b) Use an invariant principle to solve the equation.

(a) If they grew the same amount each year, what is it that stays the same? The difference in their heights stays the same. So put the difference in their heights last year on one side of the equation and the difference of their heights this year on the other side:

$$\underset{\text{last year}}{146 - 138} \;=\; \underset{\text{this year}}{151 - x}$$

(b) 146 increases by 5 to become 151, so 138 must increase by 5 to become x: $x = 138 + 5 = 143$ cm.

EXERCISE 1.9

DEVELOP YOUR SKILL

Use an invariant principle to write equivalent sums or differences that are easier to compute.

1. $142 - 97$ **2.** $176 - 58$ **3.** $94 + 178$

4. $243 + 29$ **5.** $164 - 49$ **6.** $175 - 38$

Solve each equation by using an invariant principle.

7. $434 - 295 = x - 293$ **8.** $912 - x = 916 - 567$

9. $736 + 572 = 726 + x$ **10.** $x + 183 = 463 + 283$

11. Last year at the county fair, there were 413 adults and 349 children. The same amount of people came to the fair this year, but there were 425 adults. How many children came to the fair this year?

    (a) Let x represent the number of children who came to the fair this year. Write an equation with a sum on each side that describes this problem.

    (b) Use an invariant principle to solve the equation in part (a).

MAINTAIN YOUR SKILL

Show that each statement is true by writing the equation required by its definition.
[1.2, 1.3, 1.4]

12. 22 is an even number   **13.** $23 > 19$ **14.** $17 \leq 25$

15. $15 \geq 15$ **16.** $28 > 21$ **17.** $6 < 20$

18. If the discount d is subtracted from the marked price M, this difference is the selling price S. Let $M - d = S$. Use the definition of subtraction to write an equation that tells how to compute the marked price M for a desired discount d and selling price S. [1.5]

Write all other possible equivalent forms that result from using only the commutative principle for addition. [1.7]

19. $3 + (y - 5)$ **20.** $(6 + a) - w$ **21.** $a - (b + c)$

22. $r + (s - t)$ **23.** $(x - y) + z$ **24.** $(d + e) - f$

CHAPTER 1 REVIEW

1. What, by definition, is the sum of 7 and 9? [1.1]

2. What, by definition, is the sum of n and 3? [1.1]

Show that each statement is true by writing the equation required by its definition.
[1.2, 1.3, 1.4]

3. 26 is an even number.   **4.** $7 \leq 9$

5. $7 < 10$ **6.** $21 \leq 21$

7. $15 > 11$ **8.** $30 \geq 14$

Use the definition of subtraction to change each subtraction equation to the corresponding addition equation and each addition equation to the corresponding subtraction equation. [1.5]

9. $28 - 15 = 13$ **10.** $19 + x = 27$

11. $m + 16 = 34$ **12.** $24 - n = y$

Compute the basic numeral for each expression that is defined as a counting number. Otherwise, write "not defined." [1.6]

13. $15 - 3 + 8$ **14.** $(15 - 12) - (8 - 7)$

15. $14 - 3 + 2$ **16.** $3 - 7 + 4$

17. $18 - 9 - 3$ **18.** $4 + 3 - 5$

19. $12 - 8 - 6$ **20.** $13 - 7 + 1$

21. $(9 - 5) + 11$ **22.** $15 - (2 + 9)$

Write all other possible equivalent forms that result from using only the commutative principle for addition. If there are none, write "none." [1.7]

23. $(x + 10) - y$ **24.** $x + (8 - y)$

25. $15 - (x + y)$ **26.** $20 - (x - y)$

Solve each equation by using an invariant principle. [1.8, 1.9]

27. $543 + 376 = 540 + x$ **28.** $x - 168 = 543 - 368$

29. $441 - x = 541 - 374$ **30.** $825 + x = 835 + 146$

31. $398 + 279 = y + 289$ **32.** $426 - 178 = 526 - n$

Chapter 2 – Increases and Decreases

2.1 – ADDITION AND SUBTRACTION OPERATORS

OBJECTIVE

1. See addition and subtraction as operators.

There are two ways of thinking about the sum of two numbers. The first way sees *two* numbers combined to produce a third number. This is the way we viewed addition in Lesson 1.1. For example, the sum 5 + 3 is obtained by starting with two numbers, namely, 5 and 3. When they are added together, we get their sum 5 + 3. We can imagine 5 and 3 as input to an "Adding Machine" that has 5 + 3 as the output.

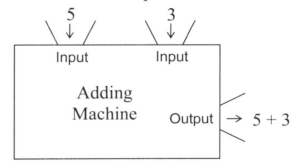

The second way of thinking about the sum 5 + 3 is to see the first number, 5, changing into the sum 5 + 3 by joining an increase of 3. We can imagine the number 5 as input to an "Increase of 3 Machine" that has 5 + 3 as the output.

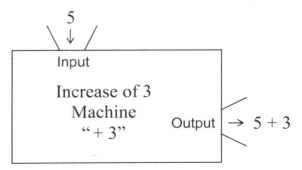

37

This second way shows one number being "operated on" or changed into another number. It emphasizes the *change* involved in forming a sum, and it frequently provides a better model for the process. The number we begin with before the operation, in this case 5, is called the **operand**. The symbol $+ 3$ is the operation we will perform, or how we will change the number. We call this the **operator**. It represents the change: an **increase** of 3. Finally, what we end up with, $5 + 3$, is the **transform**.

The number before the change $\Rightarrow$ **operand**

The change $\Rightarrow$ **operator**

The number after the change $\Rightarrow$ **transform**

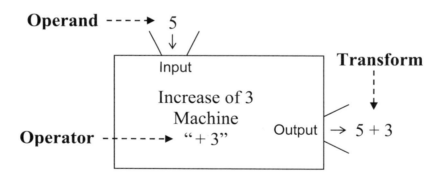

Study Tip:

Addition operators can be applied on the right or on the left. Subtraction operators can only be applied on the right.

In analyzing the sum $5 + 3$, we can also think of 3 as the operand and view $5 +$ as an addition operator applied to the left side of 3. In general, increases can be represented by **right operators** of the form $+ n$ and **left operators** of the form $n +$.

Since subtraction is not commutative, an expression such as $5 - 3$ must be computed by starting with 5 and subtracting 3 from it. So **decreases** are only represented by right operators of the form $- n$.

> ### Example 1   Name the Operand and Operator
>
> List all the ways to name the operand and the operator in each expression.
>
>   (a) $8 + 5$ (b) $7 - x$ (c) $(k + 3) - 5$
>
> (a) The operand is 8 and the operator is $+5$, an increase of 5. Or the operand is 5 and the operator is $8+$, an increase of 8 added on the left.
>
> (b) The operand is 7 and the operator is $-x$, a decrease of x.
>
> (c) Because of the parentheses around $k + 3$, we group it together as one term. The operand is $(k + 3)$ and the operator is -5, a decrease of 5.

EXERCISE 2.1

DEVELOP YOUR SKILL

See each expression as having one operand and one operator. List all the ways to name the operand and the operator.

1. $x + 9$
2. $x - 4$
3. $x - 3$
4. $18 + x$
5. $16 - x$
6. $19 - x$
7. $(b + 7) - x$
8. $x + (c - 6)$
9. $x - (3 + c)$
10. $(h - 3) + y$
11. $(w + 5) + m$
12. $n - (10 - a)$

MAINTAIN YOUR SKILL

Use the definition of subtraction to change each subtraction equation to the corresponding addition equation, and each addition equation to the corresponding subtraction equation. [1.5]

13. $37 - 25 = c$
14. $b - 24 = 35$
15. $m + 8 = n$
16. $x + y = 32$

Compute the basic numeral for each expression that is defined as a counting number. Otherwise, write "not defined." [1.6]

17. $15 + 6 - 7$

18. $12 - 4 + 9$

19. $17 - 6 - 8$

20. $3 + 6 - 7 + 5$

21. $7 - 3 - 5 + 4$

22. $8 - 5 + 1 - 7$

23. $9 - 6 + 4 - 5$

24. Last year, a gas station charged \$2.47 for a gallon of gas, and the owner of the gas station paid \$2.38. This year, the price of gas for the owner is \$2.78. What will the price of gas be so the owner makes the same profit per gallon? [1.9]

(a) Let x represent the price of gas this year. Write an equation with a difference on each side that describes this problem.

(b) Use an invariant principle to solve the equation in part (a).

2.2 – COMPOSITION OF TWO INCREASES OR TWO DECREASES

OBJECTIVE

1. Combine two increases or two decreases and display them as vectors.

The combining of two operators is called the **composition** of operators. We make this composition by writing the operators in succession. For example, the operator $+5+3$ is the composite of $+5$ and $+3$. When $+5+3$ is applied to the operand 4, the transform is the same as when the operator $+8$ is applied to 4:

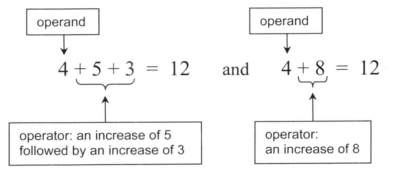

We say that two operators are **equivalent** if they give the same transform when joined to an operand. So the operator $+5+3$ is equivalent to the operator $+8$. The form $+8$ is called the **basic** operator for this combined change.

Example 1   **Find a Combined Change**

The table below records the temperature at three different times. Describe the temperature changes from 8 a.m. to 10 a.m., from 10 a.m. to noon, and the combined change from 8 a.m. to noon.

Time	Temp.
8 a.m.	70°
10 a.m.	75°
noon	78°

The change from 8 a.m. to 10 a.m. is an increase of 5, or + 5. The change from 10 a.m. to noon is an increase of 3, or + 3. The combined change from 8 a.m. to noon is an increase of 8, or + 8.

Definition:

A **vector** is a line segment that has direction and length.

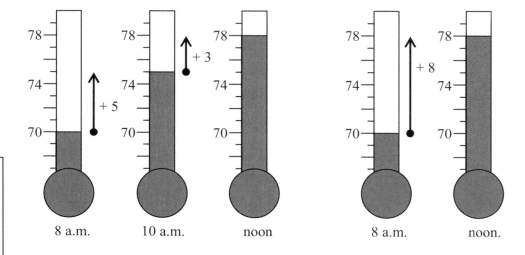

In the diagram above, we have labeled the increases of Example 1 with arrows (or **vectors**) that show the amount of the change. We can simplify the drawings by omitting the thermometers and drawing the vectors horizontally below a number line that represents the temperature. We separate the two individual increases from the composite by a broken line.

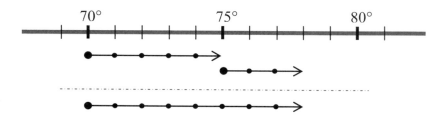

If we are only interested in the changes, we may omit the number line at the top and obtain the following vector model for the composition of the two operators $+5$ and $+3$.

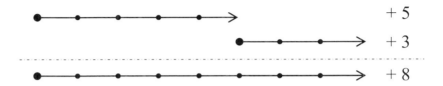

We note that the increases are modeled by vectors that point to the right. We think of them as acting along a line, but we offset the vectors so that we can see clearly where they start and stop. The second increase $+3$ begins at the point where the first increase $+5$ ends. And the composite change $+5+3$ (or $+8$) begins at the point where $+5$ begins and ends at the point where $+3$ ends.

This gives us the following operator equation

$$+5+3 \ = \ +(5+3) \ = \ +8.$$

It is also true that $+3+5 \ = \ +8$. So the two increases can be combined in either order and have the same result.

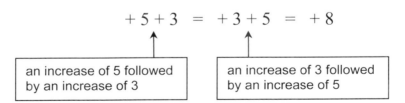

$$+5+3 \ = \ +3+5 \ = \ +8$$

an increase of 5 followed by an increase of 3

an increase of 3 followed by an increase of 5

Here is the general rule for the composition of two increases:

The operator equation

$+5+3=+8$

is read as

"An increase of 5 followed by an increase of 3 is equivalent to an increase of 8."

<u>Case 1</u> An increase of a and an increase of b.

The composite of an increase of a and an increase of b is equivalent to an increase whose magnitude is $a + b$ (or $b + a$). The operator equations are

$$+ a + b \ = \ + (a + b) \ = \ + (b + a)$$

$$+ b + a \ = \ + (b + a) \ = \ + (a + b)$$

Example 2   Find a Combined Change

The table below records the temperature at three different times. Describe the temperature changes from 4 p.m. to 6 p.m., from 6 p.m. to 8 p.m., and the combined change from 4 p.m. to 8 p.m.

Time	Temp.
4 p.m.	79°
6 p.m.	74°
8 p.m.	71°

The change from 4 p.m. to 6 p.m. is a decrease of 5, or -5. The change from 6 p.m. to 8 p.m. is a decrease of 3, or -3. The combined change from 4 p.m. to 8 p.m. is a decrease of 8, or -8.

The vector model for the changes in Example 2 is drawn with the vectors pointing to the left, since they are decreases.

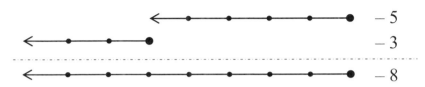

The operator equation for these changes is given by

$$-5 - 3 \ = \ -(5 + 3) \ = \ -8.$$

| a decrease of 5 followed by a decrease of 3 | a decrease of the sum 5 + 3 | a decrease of 8 |

Since it is also true that $-3 - 5 = -8$, we have the following general rule for the composition of two decreases:

<u>Case 2</u> A decrease of a and a decrease of b.

The composite of a decrease of a and a decrease of b is equivalent to a decrease whose magnitude is $a + b$ (or $b + a$). The operator equations are

$$-a - b \ = \ -(a + b) \ = \ -(b + a)$$

$$-b - a \ = \ -(b + a) \ = \ -(a + b)$$

We may summarize the composition of two increases or two decreases as follows:

Key Point

The composition of two changes of the same kind produces a larger change of the same kind. The magnitude of the composite change is the sum of the magnitudes of the individual changes.

Study Tip:

The answer to part (a) must be + 12 and not just 12. We are composing operators, not adding numbers. Our answer must be the operator + 12 and not the number 12.

Example 3   Compute the Basic Form

Compute the basic form for each composite operator.

   (a) $+5+7$ (b) $-6-9$ (c) $-3-8-2$

(a) The two increases combine to give a larger increase:
$$+5+7 \; = \; +(5+7) \; = \; +12$$

(b) The two decreases combine to give a larger decrease:
$$-6-9 \; = \; -(6+9) \; = \; -15$$

(c) This time we have three operators to combine. We can do this in either order by grouping the first two or the last two. Combining the first two operators we have
$$-3-8-2 \; = \; -(3+8)-2$$
$$= \; -11-2 \; = \; -(11+2) \; = \; -13.$$

Combining the last two operators we have
$$-3-8-2 \; = \; -3-(8+2)$$
$$= \; -3-10 \; = \; -(3+10) \; = \; -13.$$

Or we can combine all three decreases together:
$$-3-8-2 \; = \; -(3+8+2) \; = \; -13$$

EXERCISE 2.2

DEVELOP YOUR SKILL

In Exercises 1 – 4, a series of three temperatures is given. In each exercise find the following: (a) the change from the first to the second temperature; (b) the change from the second to the third temperature; (c) the composite change from the first to the third temperature; and (d) draw a vector diagram to illustrate your answers to parts (a), (b) and (c).

1. 70, 76, 80
2. 72, 75, 81
3. 85, 80, 76
4. 82, 79, 73

Construct a vector model for the composition of the operators.

5. $+3+4$
6. $+5+1$
7. $-4-5$
8. $-2-3$
9. $+2+5+1$
10. $-4-1-3$

Compute the basic form for each composite operator.

11. $+2+7$
12. $+5+9$
13. $-8-2-3$
14. $-1-8-3$
15. $+5+3+6$
16. $-2-5-1$

MAINTAIN YOUR SKILL

Show that each statement is true by writing the equation required by its definition. [1.2, 1.3, 1.4]

17. 62 is an even number.
18. $26 \geq 18$
19. $23 < 57$
20. $30 > 21$

Use the definition of subtraction to change each subtraction equation to the corresponding addition equation, and each addition equation to the corresponding subtraction equation. [1.5]

21. $x - 31 = 46$
22. $35 + w = 48$
23. $y + 7 = m$
24. $C - 8 = A$

25. A grandmother and her granddaughter have the same birthday. If the grandmother was 66 when her granddaughter was 18, how old was the grandmother when her granddaughter was 14? [1.9]

(a) Let a represent the grandmother's age. Write an equation with a difference on each side that describes this problem.

(b) Use an invariant principle to solve the equation in part (a).

2.3 – INSERTING PARENTHESES WITH LIKE CHANGES

OBJECTIVE

1. Combine two like changes into a composite change.

The last lesson showed how to combine two increases or two decreases: the result is a larger change of the same kind. In this lesson we practice this process without doing the final computation.

Example 1   Insert Parentheses

Each expression is an operand followed by two operators. Do not compute and do not change the operand, but use parentheses to write an equivalent expression with the composite operator. When possible, give an alternate form.

(a) $20 + 5 + 9$ (b) $24 - 7 - 3$

(a)   The two increases combine to give a larger increase. We want to add their magnitudes. This can be done in either order: $20 + (5 + 9)$ or $20 + (9 + 5)$.

(b)   The two decreases combine to give a larger decrease. Again, we want to add their magnitudes. This can be done in either order: $24 - (7 + 3)$ or $24 - (3 + 7)$.

When the changes involve variables, the process is the same: two increases combine to give a larger increase. Two decreases combine to give a larger decrease. In both cases we add their magnitudes.

Example 2   Insert Parentheses

Each expression is an operand followed by two operators. Do not compute and do not change the operand, but use parentheses to write an equivalent expression with the composite operator. Assume that all differences are defined. When possible, give an alternate form.

   (a) $15 + x + y$ (b) $15 - x - y$

(a) The two increases combine to give a larger increase:
$$15 + (x + y) \quad \text{or} \quad 15 + (y + x).$$

(b) The two decreases combine to give a larger decrease:
$$15 - (x + y) \quad \text{or} \quad 15 - (y + x).$$

EXERCISE 2.3

DEVELOP YOUR SKILL

Each expression is an operand followed by two operators. Do not compute and do not change the operand, but use parentheses to write an equivalent expression with the composite operator. When possible, give an alternate form.

  1. $18 - 3 - 4$ **2.** $15 + 2 + 7$

  3. $17 + 2 + 8$ **4.** $12 - 5 - 3$

Each expression is an operand followed by two operators. Do not compute and do not change the operand, but use parentheses to write an equivalent expression with the composite operator. Assume that all differences are defined. When possible, give an alternate form.

5. $25 - m - n$

6. $17 + m + n$

7. $x - a - 5$

8. $y + 4 + b$

9. $x + n + 3$

10. $a - 6 - c$

MAINTAIN YOUR SKILL

11. There were seats for 1,200 people in the auditorium. If 743 seats were taken, how many seats were empty? [1.5]

Compute the basic numeral for each expression that is defined as a counting number. Otherwise, write "not defined." [1.6]

12. $4 - 8 + 7$

13. $14 - 8 - 3$

14. $3 + 6 - 5$

15. $10 - 2 + 3$

16. $14 - 8 + 2$

17. $6 - 5 - 2$

18. $15 + 2 - 8 + 6$

19. $12 - 4 - 5 + 6$

20. $13 - (5 + 2) + 3$

21. $17 - (16 - 8 + 2)$

Solve each equation by using an invariant principle. [1.9]

22. $437 - 288 = 337 - x$

23. $138 + 265 = 158 + x$

24. $623 + x = 653 + 549$

25. $534 - x = 504 - 258$

2.4 – COMPOSITION OF AN INCREASE AND A DECREASE

OBJECTIVE

1. Combine an increase with a decrease and display them as vectors.

In Lesson 2.2 we saw how to combine two increases or two decreases: the result is a larger change of the same kind. When combining an increase with a decrease, the two changes work against each other, so the overall composite change will be smaller than either of the individual changes.

Example 1 Find a Combined Change

The table below records the temperature at three different times. Describe the temperature changes from noon to 4 p.m., from 4 p.m. to 8 p.m., and the combined change from noon to 8 p.m.

Time	Temp.
noon.	70°
4 p.m.	78°
8 p.m.	73°

The change from noon to 4 p.m. is an increase of 8, or + 8. The change from 4 p.m. to 8 p.m. is a decrease of 5, or – 5. The combined change from noon to 8 p.m. is an increase of 3, or + 3.

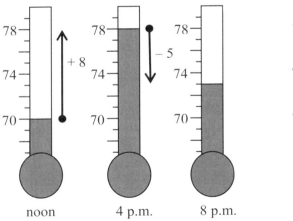
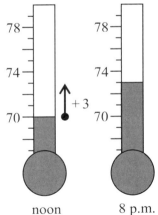

Drawing the vectors horizontally we have the following model for the composition of these changes:

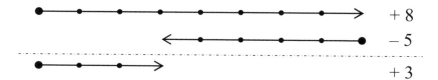

We see that the two changes are working against each other. Since the increase is larger, the overall composite change is an increase. The magnitude of the composite change is their difference.

When we focus our attention on the operators involved, we obtain the operator equation

$$+ 8 - 5 \quad = \quad + (8 - 5) \quad = \quad + 3.$$

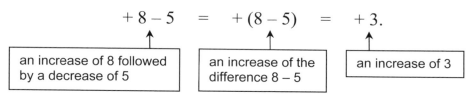

We would have obtained the same composite change if the decrease of 5 had come first and then the increase of 8:

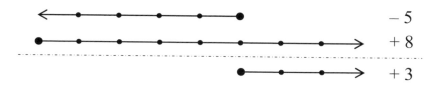

Here is the corresponding operator equation:

$$- 5 + 8 \quad = \quad + (8 - 5) \quad = \quad + 3.$$

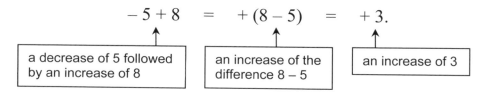

We have the following general rule for the composition of an increase and a decrease, when the increase is larger:

Study Tip:

It does not matter which change comes first. The increase is larger than the decrease, so the combined change is an increase.

<u>Case 3</u> An increase of a and a decrease of b, where $a \geq b$.

If $a \geq b$, the composite of an increase of a and a decrease of b is equivalent to an increase whose magnitude is $a - b$. The operator equations are

$$+ a - b \; = \; + (a - b)$$

$$- b + a \; = \; + (a - b)$$

Example 2   **Find a Combined Change**

The table below records the temperature at three different times. Describe the temperature changes from noon to 4 p.m., from 4 p.m. to 8 p.m., and the combined change from noon to 8 p.m.

Time	Temp.
Noon	75°
4 p.m.	78°
8 p.m.	70°

The change from noon to 4 p.m. is an increase of 3, or $+ 3$. The change from 4 p.m. to 8 p.m. is a decrease of 8, or $- 8$. The combined change from noon to 8 p.m. is a decrease of 5, or $- 5$.

This time the decrease is larger than the increase, so the overall composite change is a decrease. Here is the vector model for the composition of the operators:

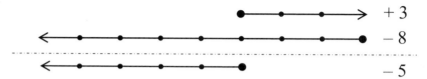

Once again, the two changes could occur in either order and produce the same result. Here is the general rule for the composition of an increase and a decrease, when the decrease is larger:

<u>Case 4</u> An increase of a and a decrease of b, where $b \geq a$.

If $b \geq a$, the composite of an increase of a and a decrease of b is equivalent to a decrease whose magnitude is $b - a$. The operator equations are

$$+ a - b = -(b - a)$$
$$- b + a = -(b - a)$$

Here are the steps to follow in combining an increase and a decrease:

1. Decide which is larger. It will determine the kind of the composite change. If the larger change is an increase, then the composite change will be an increase. If the larger change is a decrease, then the composite change will be a decrease.

2. To find the magnitude of the composite change, take the difference in the magnitudes of the two individual changes. Subtract the smaller magnitude from the larger.

Study Tip:

The answer must be +4 and not just 4. We are composing operators and not adding and subtracting numbers. Our answer must be the operator +4 and not the number 4.

Example 3   Compute the Basic Form

Compute the basic form for the composite operator

$$- 5 + 9.$$

Since the increase of 9 is larger than the decrease of 5, the combined change will be an increase. Since the two operators are working against each other, the size of the change is the difference in their magnitudes, $9 - 5$. That is,

$$- 5 + 9 \;=\; +(9 - 5) \;=\; +4.$$

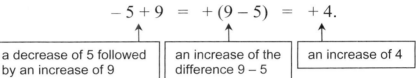

| a decrease of 5 followed by an increase of 9 | an increase of the difference $9 - 5$ | an increase of 4 |

Example 4   Compute the Basic Form

Compute the basic form for the composite operator

$$+ 3 - 10.$$

This time the decrease of 10 is larger than the increase of 3, so the combined change will be a decrease. The two operators are working against each other, so the size of the change is the difference in their magnitudes, $10 - 3$. That is,

$$+ 3 - 10 \;=\; -(10 - 3) \;=\; -7.$$

| an increase of 3 followed by a decrease of 10 | a decrease of the difference $10 - 3$ | a decrease of 7 |

In summary, we have the following rules for the composition of addition and subtraction operators (increases and decreases):

Key Point

1. Like changes help each other to give a larger composite change of the same kind. Add their magnitudes.

2. Opposite changes work against each other. The larger change will dominate the smaller and determine the kind of composite change. Subtract the smaller magnitude from the larger.

Example 5   Compute the Basic Form

Compute the basic form for the composite operator
$$-7 + 3 - 5.$$

This time there are three operators to combine. We can do this from left to right or by combining the two decreases together and then combining with the increase.

<u>From left to right</u>

$$-7 + 3 - 5 = -(7 - 3) - 5$$
The composition of a decrease of 7 and an increase of 3 is a decrease of their difference 7 − 3.

$$= -4 - 5$$
Compute 7 − 3 = 4.

$$= -(4 + 5)$$
We saw in Lesson 2.2 that the composition of two decreases is a decrease of their sum.

$$= -9$$
Compute 4 + 5 = 9.

First combine the decreases

$$-7 + 3 - 5 = -7 - 5 + 3$$

Interchange the increase of 3 and the decrease of 5.

$$= -(7 + 5) + 3$$

The composition of two decreases is a decrease of their sum.

$$= -12 + 3$$

Compute 7 + 5 = 12.

$$= -9$$

The composition of a decrease of 12 and an increase of 3 is a decrease of 9.

EXERCISE 2.4

DEVELOP YOUR SKILL

In Exercises 1 – 4, a series of three temperatures is given. In each exercise find the following: (a) the change from the first to the second temperature; (b) the change from the second to the third temperature; (c) the composite change from the first to the third temperature; and (d) draw a vector diagram to illustrate your answers to parts (a), (b) and (c).

1. 70, 76, 72
2. 72, 77, 68
3. 80, 72, 76
4. 78, 75, 80

Construct a vector model for the composition of the operators.

5. $+2 + 3$
6. $+4 - 6$
7. $-2 + 7$
8. $-4 - 2$
9. $-3 + 2$
10. $+6 - 1$

Compute the basic form for each composite operator.

11. $+5 - 8$
12. $-12 + 5$
13. $+6 + 11$
14. $-13 - 7$
15. $+18 - 6$
16. $+7 - 15$
17. $-17 + 12 - 9$
18. $+5 - 8 + 14$

19. The temperature rose 7° in the first hour, rose 9° in the second hour, and fell 5° in the third hour. Construct an operator equation to model the changes over this three hour period. Put a sequence of operators on the left side of the equation and the basic form of the composite operator on the right side.

MAINTAIN YOUR SKILL

Use the definition of subtraction to change each subtraction equation to the corresponding addition equation, and each addition equation to the corresponding subtraction equation. [1.5]

20. $25 + 17 = 42$

21. $50 - 31 = 19$

22. $36 - x = 18$

23. $y + 16 = 35$

State whether each equation illustrates the commutative principle or the associative principle. [1.7]

24. $(5 + 3) + 7 = 7 + (5 + 3)$

25. $(4 + 6) + 2 = 4 + (6 + 2)$

2.5 – INSERTING PARENTHESES WITH UNLIKE CHANGES

OBJECTIVE

1. Combine two changes into a composite change.

In the last few lessons we have seen how increases and decreases interact to form a composite change:

> 1. Like changes help each other to give a larger composite change of the same kind. Add their magnitudes.
>
> $$+4 + 5 = +9 \quad \text{and} \quad -4 - 5 = -9$$
>
> 2. Opposite changes work against each other. The larger change will dominate the smaller and determine the kind of composite change. Subtract the smaller magnitude from the larger.
>
> $$+3 - 7 = -4 \quad \text{and} \quad -3 + 7 = +4$$

It is important to gain confidence in forming these combinations. In Lesson 2.3 we practiced this when the changes were of the same kind. Now we extend this to include the case where the changes are opposites. As in Lesson 2.3, we will stop before doing the final computation. This will help prepare us for algebra, where the use of letters prevents us from relying on the numerical values.

Example 1   Insert Parentheses

Each expression is an operand followed by two operators. Do not compute and do not change the operand, but use parentheses to write an equivalent expression with the composite operator. When possible, give an alternate form.

(a) $21 - 9 - 5$ (b) $25 + 3 - 6$

(c) $25 - 6 + 3$ (d) $20 - 4 + 7$

(a) $21 - 9 - 5$

The two decreases combine to give a larger decrease. We want to add their magnitudes. This can be done in either order: $21 - (9 + 5)$ or $21 - (5 + 9)$.

(b) $25 + 3 - 6$

The increase and the decrease are working against each other and the decrease is stronger. This means the composite change is a decrease. The size of the decrease is the difference in the magnitudes: $25 - (6 - 3)$.

(c) $25 - 6 + 3$

This is the same as (b). It doesn't matter which change comes first, only which change is stronger. We still get a decrease of the difference: $25 - (6 - 3)$.

Study Tip:

With like changes there will be two possible answers because there is a sum inside the parentheses.

With opposite changes, there will be only one answer.

(d) $\qquad\qquad 20 - 4 + 7$

This time the increase is stronger, so the combined change is an increase. Because they are working against each other, the magnitude is the difference: $20 + (7 - 4)$.

The process of inserting parentheses with variables is basically the same as with numbers. Are the two changes the same? If so, then the composite is a larger change of the same kind. If the two changes are not the same, then the larger determines the kind of a composite change, and the magnitude is the difference.

Example 2   Insert Parentheses

Each expression is an operand followed by two operators. Do not change the operand, but use parentheses to write an equivalent expression with the composite operator. Assume that $x \geq y$ and that all differences are defined. When possible, give an alternate form.

(a) $24 - y + x$ $\qquad\qquad$ (b) $24 + x - y$

(c) $24 - x - y$ $\qquad\qquad$ (d) $24 + x + y$

(e) $24 - x + y$ $\qquad\qquad$ (f) $24 + y - x$

(a) $\qquad\qquad 24 - y + x$

The changes are working against each other and the increase is stronger. Subtract the smaller magnitude from the larger: $24 + (x - y)$.

(b) $\qquad\qquad 24 + x - y$

This is the same as part (a). It doesn't matter which change comes first: $24 + (x - y)$.

(c) $\qquad$ $24 - x - y$

The two decreases combine to give a larger decrease:
$24 - (x + y)$ or $24 - (y + x)$.

(d) $\qquad$ $24 + x + y$

The two increases combine to give a larger increase:
$24 + (x + y)$ or $24 + (y + x)$.

(e) $\qquad$ $24 - x + y$

The changes are working against each other and the decrease is stronger. Subtract the smaller magnitude from the larger: $24 - (x - y)$.

(f) $\qquad$ $24 + y - x$

This is the same as part (e): $24 - (x - y)$.

EXERCISE 2.5

DEVELOP YOUR SKILL

Each expression is an operand followed by two operators. Do not compute and do not change the operand, but use parentheses to write an equivalent expression with the composite operator. When possible, give an alternate form.

1. $15 - 3 - 7$
2. $12 + 2 + 6$
3. $15 - 7 + 3$
4. $12 + 6 - 2$
5. $15 + 7 - 3$
6. $12 + 2 - 6$
7. $15 + 3 + 7$
8. $12 - 6 - 2$

Each expression is an operand followed by two operators. Do not change the operand, but use parentheses to write an equivalent expression with the composite operator. Assume that $m \geq n$ and that all differences are defined. When possible, give an alternate form.

9. $30 - m - n$
10. $30 - m + n$

11. $30 - n + m$

13. $30 + n - m$

12. $30 + n + m$

14. $30 + m - n$

MAINTAIN YOUR SKILL

Show that each statement is true by writing the equation required by its definition.
[1.3, 1.4]

15. $2 < 15$

16. $24 > 18$

17. $25 \leq 29$

Use the definition of subtraction to change each subtraction equation to the corresponding addition equation, and each addition equation to the corresponding subtraction equation. [1.5]

18. $x - n = 27$

20. $y + d = 15$

19. $20 + a = b$

21. $m - 13 = k$

Construct a vector model for the composition of the operators. [2.3]

22. $+ 7 - 3$

24. $- 5 + 3$

23. $+ 2 - 8$

25. $- 2 - 5$

2.6 – INVERSE OPERATORS

OBJECTIVE

1. Identify and use inverse operators.

The operators $+ 0$, $0 +$, and $- 0$ represent no change; they are called the **identity** operators for increases and decreases. The operators $+ 3$ (an increase of 3) and $- 3$ (a decrease of 3) give changes of the same magnitude (size), but they are opposite in kind. In general, $+ n$ and $- n$ are said to be **inverse** to each other.

If two inverse operators such as $+ 3$ and $- 3$ are used in immediate succession, their changes "undo" or "cancel" each other, so that the final result is no change at all.

For example, suppose Alice gives Laura 3 dollars and Laura gives Judy 3 dollars. The overall change for Laura is $+0$. She received 3 dollars and gave 3 dollars away. We could write this in an operator equation like this:

$$+3 - 3 = +0$$

In general, an increase $+n$ followed by a decrease $-n$ is equivalent to no change, or $+0$.

$$+n - n = +0$$

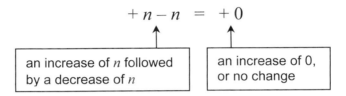

Similarly, suppose Laura gave Alice 5 dollars and then Judy gave Laura 5 dollars. Again, the overall change for Laura is $+0$. She gave away 5 dollars and then received 5 dollars:

$$-5 + 5 = +0$$

In general, a decrease followed by an increase of the same magnitude is equivalent to no change when the operand is large enough for the expression to be defined.

$$-n + n = +0$$

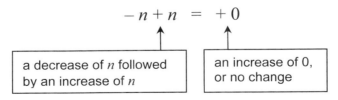

Example 1   Find the Inverse of an Operator

Write the inverse of each operator.

   (a) $+9$ (b) -2

(a) The inverse of an increase of 9 is a decrease of 9: -9.

(b) The inverse of a decrease of 2 is an increase of 2: $+2$.

Example 2   Find the Missing Operator

Find the missing operator in each equation.

   (a) $+5$ $\boxed{?}$ $= +0$ (b) -3 $\boxed{?}$ $= +0$

   (c) $\boxed{?}$ $+8 = +0$ (d) $\boxed{?}$ $-6 = +0$

In each case the missing operator is the inverse of the operator that is with it.

   (a) -5 (b) $+3$ (c) -8 (d) $+6$

EXERCISE 2.6

DEVELOP YOUR SKILL

Write the inverse of each operator.

   1. $+8$ **2.** $+5$ **3.** -7 **4.** -4

Find the missing operator in each equation.

   5. $+6$ $\boxed{?}$ $= +0$ **6.** -4 $\boxed{?}$ $= +0$

   7. $\boxed{?}$ $-7 = +0$ **8.** $\boxed{?}$ $+9 = +0$

MAINTAIN YOUR SKILL

See each expression as having one operand and one operator. List all the ways to name the operand and the operator. [2.1]

9. $y + (n - 6)$

10. $x - (m - 5)$

11. $(b + 2) - a$

12. $m + (7 + c)$

Compute the basic form for each composite operator. [2.4]

13. $+ 6 - 10$

14. $- 11 + 7$

15. $+ 5 + 13$

16. $- 14 - 5$

17. $+ 15 - 7$

18. $+ 8 - 13$

Each expression is an operand followed by two operators. Do not compute and do not change the operand, but use parentheses to write an equivalent expression with the composite operator. When possible, give an alternate form. [2.5]

19. $16 + 8 - 3$

20. $16 + 3 - 14$

21. $27 - 9 - 5$

22. $17 - 8 + 3$

23. $13 + 8 + 6$

24. $31 - 6 + 9$

25. On Monday, the value of WXY stock rose 5 cents. On Tuesday, it fell 8 cents. Construct an operator equation to model the changes in the value of WXY stock over these two days. Put two operators on the left side of the equation and the basic form of the composite operator on the right side. [2.4]

CHAPTER 2 REVIEW

See each expression as having one operand and one operator. List all the ways to name the operand and the operator. [2.1]

1. $x + 9$

2. $y - 5$

3. $12 + y$

4. $13 - m$

5. $(x + 5) - n$

6. $25 - (x - 6)$

Compute the basic form for each composite operator. [2.2, 2.4]

7. $-3 + 12$

8. $+4 - 10$

9. $-8 - 5$

10. $+12 - 5$

11. $-11 + 7$

12. $+9 + 14$

13. The thermometer reading rose 12 degrees from noon to 6 p.m., and then fell 8 degrees from 6 p.m. to midnight. Compute the combined change in temperature for this 12-hour period. [2.4]

Construct a vector model for the composition of the operators. [2.2, 2.4]

14. $-3 + 7$

15. $-3 - 4$

Each expression is an operand followed by two operators. Do not compute and do not change the operand, but use parentheses to write an equivalent expression with the composite operator. When possible, give an alternate form. [2.3, 2.5]

16. $4 + 5 + 6$

17. $5 + 6 - 7$

18. $26 - 8 - 3$

19. $12 - 8 + 3$

20. $21 + 10 - 7$

21. $16 - 4 + 9$

22. $12 + 4 - 11$

23. $27 - 12 + 5$

Each expression is an operand followed by two operators. Do not change the operand, but use parentheses to write an equivalent expression with the composite operator. Assume that $c \geq d$ and that all differences are defined. When possible, give an alternate form. [2.3, 2.5]

24. $25 - c + d$

25. $25 + d + c$

26. $25 - c - d$

27. $25 + d - c$

28. $25 + c - d$

29. $25 - d + c$

Write the inverse of each operator. [2.6]

30. $+16$

31. -23

Chapter 3 – Products and Quotients

3.1 – MULTIPLICATION

OBJECTIVES

1. Express multiplication as repeated addition.

2. Model multiplication by comparing line segments.

When an increase is repeatedly applied to 0, this sequence of increases can be abbreviated by multiplication.

If x and y are counting numbers, then multiplication assigns to x and y a unique third counting number $x \cdot y$, called the **product** of x and y.

In the product $x \cdot y$, we may think of x as counting the number of times that the increase $+y$ is applied to 0.

Example 1		**An Operator Model for Multiplication**		
Product Form	Number of Increases	Operator Model	Description	Basic Numeral
$0 \cdot 5$	none	0	5 is added to 0, 0 times	0
$1 \cdot 5$	one	$0 + 5$	5 is added to 0, 1 time	5
$2 \cdot 5$	two	$0 + 5 + 5$	5 is added to 0, 2 times	10
$3 \cdot 5$	three	$0 + 5 + 5 + 5$	5 is added to 0, 3 times	15

Study Tip:

3 and 5 are **factors** of 15, because $3 \cdot 5 = 15$.

The word "product," like sum and difference, has two meanings. When we multiply 3 and 5, we can compute and get a basic numeral of 15. So 15 is the product of 3 and 5. But $3 \cdot 5$ is also called the product of 3 and 5 because of its form. The numbers 3 and 5 are called **factors** of the product $3 \cdot 5$, and hence are factors of 15. In general,

$$[\text{factor}] \times [\text{factor}] = [\text{product}].$$

We can write products in several ways:

3×5	$(3)(5)$	$(3)5$	$x(y)$	xy	$3x$
$3 \cdot 5$	$(x)(y)$	$3(5)$	$(x)y$	$x \cdot y$	$3 \cdot x$

Sometimes we even use brackets when more than one product is indicated: $[(3)(5)](7)$.

Example 1 shows us the special properties that the number zero has as a factor of a product. In the product $0 \cdot x$, there is no increase applied to 0. So for each counting number x we have

$$0 \cdot x = 0.$$

On the other hand, in the product $x \cdot 0$, we are told to start with 0 and add 0 a total of x times. But no matter how many times the operator $+ 0$ is joined to 0, the result will still be 0. So for each counting number x we have

$$x \cdot 0 = 0.$$

Furthermore, if $x \cdot y = 0$, then either x or y or both must be zero. Otherwise, when x and y are both not zero, there would be at least one nonzero increase so that the final result could not be zero. This is called the Zero-Factor Property.

Zero-Factor Property

If $x \cdot y = 0$, then either $x = 0$ or $y = 0$.

Example 1 also shows us that $1 \cdot x = 0 + x = x$, and it can be used to check that $x \cdot 1 = x$, for all x. For this reason, the number 1 is called the **identity** for multiplication.

Identity Property

Study Tip:

When writing an expansion of 5 by itself, it is best to use ( )5 or 5( ) since the raised dot in · 5 is easily missed.

For all x, $1 \cdot x = x$ and $x \cdot 1 = x$.

A product such as $3 \cdot 5$ or $3(5)$ can be formed by applying the multiplication operator $3 \cdot$ or $3(\ )$ to the operand 5. Both $3 \cdot$ and $3(\ )$ are left operators.

We can also view the product $3 \cdot 5$ or $(3)5$ as the result when the right operator $\cdot 5$ or $(\ )5$ is applied to the operand 3. We see that multiplication operators, like addition operators, can be applied on the right or on the left.

If $n > 1$ and $x > 0$, then the multiplication operator $n(\ )$ applied to the operand x yields a transform that is larger than x. For this reason, multiplication operators are called **expansions.**

To obtain a graphical model for expansions, we compare the length of line segments, where the length of the second line segment results from the expansion applied to the first line segment. For example, to visualize the action of the operator $3(\ )$, we start with a line segment such as ●——● and copy it 3 times below. The length of the lower line is then the length of the upper line times 3.

It may be helpful to think of the line segment as a rubber band that is stretched to 3 times its length.

EXERCISE 3.1

DEVELOP YOUR SKILL

1. What, by definition, is the product of 4 and 9?

2. What, by definition, is the product of 6 and m?

Compute the basic numeral for each product. If no computation is possible, write "not possible."

3. $5 \cdot 8$

4. $2 \cdot 6$

5. $3 \cdot 8$

6. $7 \cdot 3$

7. $8 \cdot 4$

8. $y \cdot 5$

9. $9 \cdot 5$

10. $10 \cdot 7$

11. $6 \cdot 4$

12. $11 \cdot 6$

13. $13 \cdot x$

14. $12 \cdot 5$

15. Write an operator model for $2 \cdot 6$, as in Example 1.

16. Write an operator model for $6 \cdot 2$, as in Example 1.

Construct a graphical model for each expansion operator.

17. $4(\ )$

18. $2(\ )$

19. $6(\ )$

MAINTAIN YOUR SKILL

20. In the first hour, Jill sharpened seventy-five pencils. In the second hour she sharpened some more pencils. Altogether she sharpened one hundred twenty-seven pencils. How many pencils did Jill sharpen in the second hour? [1.5]

In Exercises 21 and 22, see each expression as having one operand and one operator. List all the ways to name the operand and the operator. [2.1]

21. $n - (7 + m)$

22. $(a - 3) + y$

Each expression is an operand followed by two operators. Do not compute and do not change the operand, but use parentheses to write an equivalent expression with the composite operator. When possible, give an alternate form. [2.3, 2.5]

23. $15 + 3 - 8$

24. $17 - 5 - 4$

25. $14 - 7 + 9$

26. $22 + 8 - 2$

3.2 – DIVISION

OBJECTIVE

1. Identify the relationship between products and quotients.

Division, like subtraction, is only a *partial* operation on the counting numbers. The division of a first number x by a second number y fails for many choices of counting numbers x and y. For example, $8 \div 3$ is not a counting number.

We all know that $8 \div 2 = 4$, but what is the multiplication equation that shows this to be true? We know that

$$8 \div 2 = 4 \quad \text{because} \quad 2 \cdot 4 = 8.$$

This relationship between division and multiplication is the pattern for our definition of division.

Definition of

Division

Let x and y be counting numbers with $y \neq 0$. Then x can be divided by y, and we write $x \div y = z$ or $\frac{x}{y} = z$, if and only if there is a unique counting number z such that

$$y \cdot z = x.$$

Example 1   **Use the Definition of Division**

Use the definition of division to change each division equation to the corresponding multiplication equation.

(a) $\dfrac{27}{9} = 3$ (b) $20 \div 5 = 4$ (c) $\dfrac{p}{q} = r$

(a) $9 \cdot 3 = 27$ (b) $5 \cdot 4 = 20$ (c) $q \cdot r = p$

Note in Example 1 that the order of the terms is important. The definition calls for only *one* answer. For example, in (a) it is also true that $3 \cdot 9 = 27$, but this does not follow directly from the definition. In the definition, the two factors y and z are in the same order left to right in the multiplication and the division equations. Since 9 is to the left of 3 in the division equation, 9 must also be to the left of 3 in the multiplication equation.

Example 2   Use the Definition of Division

Use the definition of division to change the multiplication equation to the corresponding division equation.

  (a) $7 \cdot 13 = 91$ (b) $r \cdot s = t$

(a) $\dfrac{91}{7} = 13$ (b) $\dfrac{t}{r} = s$

Once again, the order of the terms is important. The definition calls for only *one* answer. For example, in (a) it is also true that $\frac{91}{13} = 7$, but this does not follow directly from the definition.

In Lesson 3.1 we saw that 1 is the identity for multiplication:

$$\text{(a)}\ 1 \cdot x = x \quad \text{and} \quad \text{(b)}\ x \cdot 1 = x.$$

When we look at the related division equations, we obtain two more useful results:

$$\text{(a)}\ \frac{x}{1} = x \quad \text{and} \quad \text{(b) if } x \neq 0, \text{ then } \frac{x}{x} = 1.$$

The definition of division requires that every quotient be related to a certain product. Since zero dominates every product, this causes difficulties when zero is used in a quotient. Our next example illustrates the three possible types of quotients involving a zero.

Example 3   Quotients Involving Zero

(a) $\dfrac{0}{3} = 0$, since $3 \cdot 0 = 0$. The definition requires that

$\dfrac{0}{y} = z$ if and only if $y \cdot z = 0$. If $y \neq 0$, then z must be

zero. This means $\dfrac{0}{y} = 0$ for all $y \neq 0$.

(b) $\dfrac{3}{0}$ is undefined since $\dfrac{x}{0} = z$ if and only if $0 \cdot z = x$.

But there is no number when multiplied by 0 will give 3. In general, if $x \neq 0$, then there is no z such that $0 \cdot z = x$. So,

$\dfrac{x}{0}$ cannot represent a number.

(c) $\dfrac{0}{0}$ is undefined. For $\dfrac{0}{0} = z$, the definition requires that

$0 \cdot z = 0$. Since for *every* z we have $0 \cdot z = 0$, then a unique

number cannot be decided for z. Hence, $\dfrac{0}{0}$ cannot be used.

From (b) and (c) we conclude that *zero must never be a divisor.*

Study Tip:

Zero can be divided by y if y is not zero. For all $y \neq 0$ we have

$\dfrac{0}{y} = 0.$

Study Tip:

Division by zero is never possible. That is,

$\dfrac{x}{0}$

is not defined for any x.

A quotient such as $\frac{12}{3}$ can be formed by applying the division operator $\frac{(\ )}{3}$ to the operand 12. We see that the result, namely 4, is smaller than the operand 12. For this reason we refer to division operators as **contractions**. Since $\frac{x}{1} = x$ for all x, the **identity** operator $\frac{(\ )}{1}$ causes no change at all.

We are limited in using division operators now because only the counting numbers are available. For the quotient $\frac{x}{n}$ to be defined, it is required that n divides x. The parentheses are part of the operator $(\ ) \div 3$ when used in $(7 + 5) \div 3$. Parentheses are not needed for $\frac{7+5}{3}$, but when written without an operand, the form $\frac{(\ )}{3}$ is less likely to read incorrectly.

Study Tip:

It may be helpful to think of the original line segment as a stretched rubber band that has its length contracted by a factor of 3.

Example 4   A Graphical Model for Division

To visualize the action of the operator $\frac{(\ )}{3}$, we start with a line segment such as ●——————● .
Copy the length of this segment below it, divide the length into 3 equal parts, and select the first part. The length of the lower line is then the length of the upper line divided by 3. In the lower line, we show the rest of the original length before the contraction by a dotted line ●·········● to make the comparison easier.

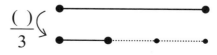

EXERCISE 3.2

DEVELOP YOUR SKILL

See each expression as having one operand and one operator. List all the ways to name the operand and the operator.

1. $7 \div n$ **2.** $x - 6$ **3.** $4m$

4. $y + 15$ **5.** $(x - 3)(8)$ **6.** $15 - (a + 2)$

For each equation, write the corresponding equation that is matched to it by the definition of division.

7. $\dfrac{68}{4} = 17$ **8.** $\dfrac{x}{7} = 21$

9. $23 \cdot 14 = 322$ **10.** $4y = 20$

11. $c \div 6 = a$ **12.** $3n = p$

13. $\dfrac{a}{b} = c + d$ **14.** $a(b - c) = d$

15. Use line segments to model the contraction operator $\dfrac{(\ \ )}{2}$.

Compute each of the following, if defined. Otherwise, write "undefined."

16. $\dfrac{0}{5}$ **17.** $\dfrac{5}{0}$ **18.** $\dfrac{0}{0}$

MAINTAIN YOUR SKILL

For each equation, write the corresponding equation that is matched to it by the definition of subtraction. [1.5]

19. $24 + 37 = 61$ **20.** $62 - 17 = 45$

21. $49 - 25 = 24$ **22.** $26 + 31 = 57$

Write all other possible equivalent forms that result from using only the commutative principle for addition. [1.7]

23. $(a + 2) - b$ **24.** $(q - p) + 5$

3.3 – Divisors and Multiples

OBJECTIVE

1. Find divisors and multiples.

The equation

$$3 \cdot 4 = 12$$

shows how the numbers 3 and 12 are related by multiplication. This relationship can be described in several equivalent ways. We say that

3 is a **factor** of 12.

3 is a **divisor** of 12.

3 **divides** 12 (or 3 **divides** into 12).

12 is a **multiple** of 3.

Study Tip:

y divides x if and only if y is a factor of x.

Example 1   Find Divisors and Multiples

For what counting numbers x are each of the following true? When possible, list the five smallest values for x.

(a) x divides 5   (b) 5 divides x (c) x divides 6

(a) Since 5 can only be factored as $1 \cdot 5$, we see that 1 and 5 are the only counting numbers that divide 5.

(b) This is asking for the multiples of 5. We have

$0 \cdot 5 = 0$, $1 \cdot 5 = 5$, $2 \cdot 5 = 10$, $3 \cdot 5 = 15$, and $4 \cdot 5 = 20$. We could keep going, but the five smallest multiples of 5 are 0, 5, 10, 15, and 20.

(c) Since 6 can only be factored as $1 \cdot 6$ or $2 \cdot 3$, we see that 1, 2, 3 and 6 are the only counting numbers that divide 6.

Example 2   **Find Divisors and Multiples**

Find all the divisors and the four smallest multiples of 12.

12 can be factored in several ways:

$$1 \cdot 12, \quad 2 \cdot 6, \quad 3 \cdot 4$$

So the divisors (factors) of 12 are 1, 2, 3, 4, 6, and 12.

The four smallest multiples of 12 are

$$0 \cdot 12 = 0, \quad 1 \cdot 12 = 12, \quad 2 \cdot 12 = 24, \quad 3 \cdot 12 = 36.$$

If y divides x, then division assigns to x and y a third number $\frac{x}{y}$ (or $x \div y$) called the **quotient** of x and y. In this form, x is called the **dividend** and y is called the **divisor**.

x is called the **dividend**.

The form $\dfrac{x}{y}$ is called a **quotient**.

y is called the **divisor**.

As we would expect, the word "quotient" has two meanings. For example, a division computation for $\frac{15}{3}$ gives the basic numeral 5 as the quotient. But $\frac{x}{y}$ and $\frac{15}{3}$ are also called quotients as a name for their form.

Example 3   Find Divisors and Multiples

For what counting numbers x are each of the following quotients defined? When possible, list the five smallest values for x.

(a) $\dfrac{6}{x}$ (b) $\dfrac{x}{6}$ (c) $\dfrac{5}{x}$

(a) The divisors of 6 are 1, 2, 3, and 6.

(b) This is asking for the multiples of 6. The five smallest are 0, 6, 12, 18, and 24.

(c) The divisors of 5 are 1 and 5.

EXERCISE 3.3

DEVELOP YOUR SKILL

Indicate for what counting numbers each statement is true. When possible, list the four smallest values of x.

1. 1 divides x **2.** x divides 1

3. 2 divides x **4.** x divides 2

5. 4 divides x **6.** x divides 4

Indicate for what counting numbers each quotient is defined. When possible, list the four smallest values.

7. $\dfrac{10}{x}$ **8.** $\dfrac{11}{x}$ **9.** $\dfrac{x}{10}$

10. $\dfrac{x}{11}$ **11.** $\dfrac{x}{15}$ **12.** $\dfrac{15}{x}$

13. Tom was 1 year old on the day his mother was 21. At that time his age divided his mother's age. At what other ages will Tom's age divide his mother's?

MAINTAIN YOUR SKILL

Each expression is an operand followed by two operators. Do not compute and do not change the operand, but use parentheses to write an equivalent expression with the composite operator. When possible, give an alternate form. [2.3, 2.5]

14. $14 + 2 - 5$ **15.** $15 - 6 - 3$

16. $16 - 4 + 7$ **17.** $17 + 9 - 5$

18. In January, Robert deposited $50 in his savings account. In February, he deposited $60. In March, he withdrew $30. Construct an operator equation to model the changes in Robert's account. Put a sequence of operators on the left side of the equation and the basic form of the composite operator on the right side. [2.4]

19. What, by definition, is the product of 6 and 3? [3.1]

20. Use line segments to model the contraction operator $\dfrac{(\ )}{4}$. [3.2]

For each equation, write the corresponding equation that is matched to it by the definition of division. [3.2]

21. $(n)(7) = w$ **22.** $3a = 5 + d$

23. $\dfrac{a - b}{c} = d$ **24.** $x \div 5 = n + 4$

3.4 – RULES OF DIVISIBILITY

OBJECTIVE

Use simple rules for divisibility.

For several numbers, there is a simple test to check if it is a divisor of another number. The most commonly used rules check for divisibility by 2, 3, 5, and 10.

Divisibility

Rules

Divisible By	Test	Example
2	The last digit is divisible by 2.	138 is divisible by 2 because the last digit is 8, and 8 is divisible by 2.
3	The sum of the digits is divisible by 3.	84 is divisible by 3 because the sum of the digits is $8 + 4 = 12$, and 12 is divisible by 3.
5	The last digit is 0 or 5.	1,285 is divisible by 5 because the last digit is 5.
10	The last digit is 0.	43,210 is divisible by 10 because the last digit is 0.

Example 1   Test for Divisibility

Determine whether 165 *is divisible by* 2, 3, 5, *or* 10.

Number	Divisible?	Reason
2	no	The last digit is 5, and 5 is not divisible by 2.
3	yes	The sum of the digits is $1 + 6 + 5 = 12$, and 12 is divisible by 3.
5	yes	The last digit is 5.
10	no	The last digit is not 0.

Example 2   Test for Divisibility

Determine whether 234 *is divisible by* 2, 3, 5, *or* 10.

Number	Divisible?	Reason
2	yes	The last digit is 4, and 4 is divisible by 2.
3	yes	The sum of the digits is $2 + 3 + 4 = 9$, and 9 is divisible by 3.
5	no	The last digit is 4, not 0 or 5.
10	no	The last digit is not 0.

The rules for divisibility are also useful in finding the factors of a number. When we find one factor, we can divide it into the number to obtain another factor.

Example 3   Find Factors

Find all the factors of 36.

We can make a chart that lists the potential factors and use division to find the other factor paired with it.

Factors of 36

Possible Divisors	Does it divide 36?	Factor Pairs
1	yes	$1 \cdot 36$
2	yes	$2 \cdot 18$
3	yes	$3 \cdot 12$

4	yes	4·9
5	no	none
6	yes	6·6

We don't have to check divisors greater than 6 in this case. For any factor greater than 6, the factor paired with it must be less than 6, and it would already appear in the list of factor pairs. Thus the factors of 36 are 1, 2, 3, 4, 6, 9, 12, 18, and 36.

EXERCISE 3.4
DEVELOP YOUR SKILL
Determine whether each number is divisible by 2, 3, 5 or 10.

1. 62   **2.** 51   **3.** 345

4. 270   **5.** 1507   **6.** 1701

List all the factors of each number.

7. 21   **8.** 28

9. 40   **10.** 45

MAINTAIN YOUR SKILL
11. There are 4 people in each car and there are 6 cars. How many people are there in all? [3.1]

Solve each equation by using an invariant principle. [1.9]

11. What is x, if $548 + 169 = 538 + x$?

12. What is x, if $824 - 276 = 844 - x$?

Compute the basic form for each composite operator. [2.4]

13. $-8 + 2$ **14.** $+7 + 5$

15. $+10 - 14$ **16.** $-5 - 11$

17. $-4 + 7 - 1$ **18.** $+6 - 4 + 9$

19. $+3 - 5 + 8$ **20.** $-7 - 4 + 2$

For each equation, write the corresponding equation that is matched to it by the definition of division. [3.2]

21. $126 \div 7 = 18$ **22.** $15 \cdot 13 = 195$

23. $(b - 5)c = 48$ **24.** $\dfrac{a + 4}{3} = h$

3.5 – ORDER OF OPERATIONS

OBJECTIVE

1. Use the order of operations.

Algebraic expressions often have a mixture of addition, subtraction, multiplication, and division operations. Special attention must be given to the order in which they are computed. Parentheses may be used, as in Lesson 1.6, to specify what is done first, what is next, and so on. Thus $3 + (4 \cdot 5)$ is computed by first multiplying 4 times 5, and then adding 3. But it is possible to use fewer parentheses and simplify the notation by adopting the following rule:

Order of Operations

> Multiplication and division are computed before addition and subtraction, unless there are instructions otherwise.

Using this rule we see that $3 + 4 \cdot 5$ is computed as $3 + (4 \cdot 5)$ by first multiplying $4 \cdot 5$ even though this does not follow the left-to-right reading order. If we want to have adding or subtracting done before multiplying or dividing, we show this by using parentheses.

> **Example 1   The Order of Operations**
>
> $3 + 4 \cdot 5 = 3 + 20 = 23$ Multiply, then add.
>
> $(3 + 4) \cdot 5 = 7 \cdot 5 = 35$ Add, then multiply.
>
> $3 \cdot (4 + 5) = 3 \cdot 9 = 27$ Add, then multiply.

Note that the bar notation for quotients has the same effect as parentheses. The bar requires that division be done last. So $\dfrac{12 + 6}{2}$ is the same as $\dfrac{(12 + 6)}{2}$, and the addition must be done first. We have

$$\frac{12 + 6}{2} = \frac{18}{2} = 9.$$

If the division sign $\div$ were to be used, then parentheses would be needed for the same order:

$$(12 + 6) \div 2 = 18 \div 2 = 9, \quad \text{but}$$

$$12 + 6 \div 2 = 12 + 3 = 15.$$

When computing with products and quotients, work left to right. We have

$$12 \div 3 \times 2 = (12 \div 3) \times 2$$
$$= 4 \times 2 = 8$$

But $12 \div (3 \times 2) = 12 \div 6 = 2.$

When more than one operation is used in an expression, the name given to the whole expression comes from the *last* operation used. So, $3 + 4 \cdot 5$ is a sum and $(3 + 4) \cdot 5$ is a product.

Study Tip:

The name of the form comes from the **last** operation used.

Example 2   Name the Form and Compute

Name each form and then compute the basic numeral.

(a) $3 + \dfrac{4 \cdot 6}{8}$ (b) $\dfrac{3+7}{3+2}$ (c) $25 - 4 \cdot 3$

(a) The last operation is addition, so this is a sum. We have

$$3 + \frac{4 \cdot 6}{8} = 3 + \frac{24}{8} = 3 + 3 = 6.$$

(b) This is a quotient. We have $\dfrac{3+7}{3+2} = \dfrac{10}{5} = 2.$

(c) This is a difference. We have $25 - 4 \cdot 3 = 25 - 12 = 13.$

EXERCISE 3.5

DEVELOP YOUR SKILL

Name each form (sum, difference, product, or quotient) and then compute the basic numeral.

1. $15 - 2 \cdot 3$

2. $4(5 + 2)$

3. $4 + 3 \cdot 6$

4. $5 \cdot 2 - 3$

5. $6 \cdot 4 \div 2$

6. $7\left(\dfrac{12}{3}\right)$

7. $3 \cdot 4 + 3 \cdot 5$

8. $3 \cdot 6 + 20 \div 4$

9. $\left(\dfrac{7+8}{3}\right)5$

10. $\dfrac{6 + 2 \cdot 5}{2}$

MAINTAIN YOUR SKILL

Compute the basic form for the composite operator. [2.4]

11. $+15-22$

12. $+7+12$

13. $-8-16$

14. $-22+9$

Each expression is an operand followed by two operators. Do not compute and do not change the operand, but use parentheses to write an equivalent expression with the composite operator. When possible, give an alternate form. [2.3, 2.5]

15. $16-3-8$

16. $11+16-19$

17. $7+14-5$

18. $12-8-2$

Construct a graphical model for each operator. [2.2, 3.1, 3.2]

19. $+5$

20. -6

21. $5(\ )$

22. $\dfrac{(\ )}{6}$

Indicate for what counting numbers each statement is true. When possible, list the four smallest values of x. [3.3]

23. 3 divides x

24. x divides 3

3.6 – PROPERTIES OF PRODUCTS AND QUOTIENTS

OBJECTIVE

1. Identify and use the commutative and associative principles for multiplication

Multiplication, like addition, is commutative. This means, for example, that $3 \cdot 5$ and $5 \cdot 3$ both represent the same number. In fact, from the definition of multiplication in Lesson 3.1, we know that $3 \cdot 5$ is 5 added to 0, 3 times:

$$3 \cdot 5 = 0 + 5 + 5 + 5 = 15$$

And $5 \cdot 3$ is 3 added to 0, 5 times:

$$5 \cdot 3 = 0 + 3 + 3 + 3 + 3 + 3 = 15$$

The **commutative principle for multiplication** says that in *any* product, the order of the two factors can be reversed:

Key Concept

If x and y are counting numbers, then $x \cdot y = y \cdot x$.

If we interchange the order in a quotient, then the result is usually not equivalent. For example,

$$\frac{8}{2} = 4 \quad \text{but} \quad \frac{2}{8} \text{ is not defined as a counting number.}$$

So, division is *not* commutative.

Example 1   Use the Commutative Principle

Use the commutative principle for multiplication to replace each expression by an equivalent expression. Do not compute.

   (a) $4 \cdot 9$ (b) $(12)(7 - 2)$

(a) $9 \cdot 4$

(b) The term $(7 - 2)$ is a factor and we interchange it with factor 12. We have $(7 - 2)(12)$.

Example 2   Use the Commutative Principles

Use the commutative principles for multiplication and for addition to replace each expression by one or more equivalent expressions. Do not compute.

   (a) $\dfrac{24}{a + 2}$ (b) $(x + b)(3)$ (c) $\dfrac{x - 5}{4}$

(a) Division is not commutative, but the addition of a and 2 in the divisor is. We get $\dfrac{24}{2 + a}$.

(b) This expression has both addition and multiplication, so there are two operations we can commute. If we commute the sum inside the parentheses we have

$$(b + x)(3)$$

If we commute the two factors $(x + b)$ and 3 we have

$$(3)(x + b).$$

And if we commute both of them we get

$$(3)(b + x).$$

(c) This expression has subtraction and division. Neither of these are commutative, so it is not possible to make any changes using a commutative principle.

Does the associative principle apply to multiplication? Let's look at an example.

$$2 \cdot (3 \cdot 5) = 2 \cdot (15) = 30$$
$$\text{and} \quad (2 \cdot 3) \cdot 5 = (6) \cdot 5 = 30$$

In the first case we multiply 3 times 5 and then multiply by 2. In the second case we multiply 2 times 3 and then multiply by 5. In both cases, we get 30. The **associative principle for multiplication** says that changing the order of the multiplying always leaves the product the same.

Key Concept

If x, y, and z are counting numbers, then

$$x \cdot (y \cdot z) = (x \cdot y) \cdot z.$$

Note that when using the commutative principle for multiplication, the order of the factors changes—they commute. When using the associative principle for multiplication, the order of the factors stays the same, but they are grouped (or associated) differently. In this case, the order of the multiplying changes.

Our next example illustrates that division is not associative. When using compound quotients, the quotient with the shorter bar is computed first. This use of the division bar reduces the need for parentheses.

Example 3   **Division is Not Associative**

$$24 \div (4 \div 2) = 24 \div 2 = 12$$

$$\text{and } (24 \div 4) \div 2 = 6 \div 2 = 3$$

Using the vertical notation and the division bar we have

$$\dfrac{24}{\dfrac{4}{2}} = \dfrac{24}{2} = 12 \quad \text{and} \quad \dfrac{\dfrac{24}{4}}{2} = \dfrac{6}{2} = 3$$

Since $12 \neq 3$, the division fails to be associative.

EXERCISE 3.6

DEVELOP YOUR SKILL

Write all other possible equivalent forms that result from using the commutative principles for addition and multiplication. If there are none, write "none." Do not compute.

1. $8(10 - 3)$

2. $\dfrac{30}{2 + 3}$

3. $\dfrac{a - b}{8}$

4. $40 - ab$

5. $\dfrac{x}{5} + 4$

6. $7 + xy$

7. $5\left(\dfrac{a}{b}\right)$

8. $\dfrac{2 \cdot 8}{y}$

9. $\dfrac{x + a}{3}$

10. $\dfrac{x}{2} - n$

11. $3y - 5$

12. $\dfrac{x}{c - d}$

In Exercises 13 and 14, state whether the commutative or the associative principle is used to make the consecutive changes of form: (a) to (b), (b) to (c), etc.

13. (a) $2 \cdot [3 \cdot (4 \cdot 5)]$
  (b) $2 \cdot [(4 \cdot 5) \cdot 3]$
  (c) $[2 \cdot (4 \cdot 5)] \cdot 3$
  (d) $[2 \cdot (5 \cdot 4)] \cdot 3$
  (e) $[(2 \cdot 5) \cdot 4] \cdot 3$
  (f) $(2 \cdot 5)(4 \cdot 3)$

14. (a) $(w \cdot z)(y \cdot x)$
  (b) $(y \cdot x)(w \cdot z)$
  (c) $y \cdot [x \cdot (w \cdot z)]$
  (d) $[x \cdot (w \cdot z)] \cdot y$
  (e) $[(x \cdot w) \cdot z] \cdot y$
  (f) $[z \cdot (x \cdot w)] \cdot y$

MAINTAIN YOUR SKILL

15. There were a total of 42 oranges in several bags. Six oranges were in each bag. How many bags were there? [3.2]

Show that each statement is true by writing the equation required by its definition. [1.3, 1.4]

16. $17 < 20$

17. $8 \le 15$

18. $13 \ge 13$

19. $19 > 11$

Compute the basic numeral for each expression that is defined as a counting number. Otherwise, write "not defined." [1.6]

20. $21 - 10 - 3$

21. $4 - 7 + 8$

22. $6 + 9 - 11$

23. $12 - 8 + 5$

Solve each equation using an invariant principle. [1.8, 1.9]

24. $235 + x = 435 + 376$

25. $x - 346 = 741 - 326$

3.7 – THE DISTRIBUTIVE PRINCIPLE

OBJECTIVE

1. Use the distributive principle.

The interactions between multiplication/division and addition/subtraction are given by the **distributive principles**. We give examples of this and then state the general rules.

Example 1   **The Distributive Principle**

(a) $2 \cdot (3 + 5) = 2 \cdot 8 = 16$

$(2 \cdot 3) + (2 \cdot 5) = 6 + 10 = 16$

We see that the *product* $2 \cdot (3 + 5)$ is equal to the *sum* $(2 \cdot 3) + (2 \cdot 5)$.

(b) $(9 - 2) \cdot 4 = 7 \cdot 4 = 28$

$(9 \cdot 4) - (2 \cdot 4) = 36 - 8 = 28$

This time the *product* $(9 - 2) \cdot 4$ is equal to the *difference* $(9 \cdot 4) - (2 \cdot 4)$.

We can model the product in Example 1(a) by using two rows of markers, with 3 white markers and 5 black markers in each row. If we add the 3 and the 5 to get 8 and then double the 8 we have a total of $2(3 + 5) = 2 \cdot 8 = 16$ markers. If we double the number of white markers to get 6, and double the number of black markers to get 10, then the total is still $2 \cdot 3 + 2 \cdot 5 = 6 + 10 = 16$.

In Example 1(a), the multiplication by 2 in the product form is distributed (spread around) and used twice in the sum form. We say that *multiplication distributes over addition*. In Example 2(b), the multiplication by 4 in the product form is distributed over the difference $9 - 2$ to give $9 \cdot 4 - 2 \cdot 4$. We say that *multiplication distributes over subtraction*.

Distributive Principles for Multiplication

If x, y and z are counting numbers, then

$$x(y + z) = xy + xz \quad \text{and} \quad (y + z)x = yx + zx.$$

Furthermore, if $y \geq z$, then

$$x(y - z) = xy - xz \quad \text{and} \quad (y - z)x = yx - zx.$$

In Example 3, we see that division also distributes over addition and subtraction.

Example 2 The Distributive Principle

$$\frac{20 + 12}{4} = \frac{32}{4} = 8 \quad \text{and} \quad \frac{20}{4} + \frac{12}{4} = 5 + 3 = 8.$$

The *quotient* $\dfrac{20 + 12}{4}$ is equal to the *sum* $\dfrac{20}{4} + \dfrac{12}{4}$.

$$\frac{20 - 12}{4} = \frac{8}{4} = 2 \quad \text{and} \quad \frac{20}{4} - \frac{12}{4} = 5 - 3 = 2$$

The *quotient* $\dfrac{20 - 12}{4}$ is equal to the *difference* $\dfrac{20}{4} - \dfrac{12}{4}$.

Distributive Principles for Division

> If x, y and z are counting numbers with z dividing both x and y, then
>
> $$\frac{x+y}{z} = \frac{x}{z} + \frac{y}{z}.$$
>
> Furthermore, if $x \geq y$, then
>
> $$\frac{x-y}{z} = \frac{x}{z} - \frac{y}{z}.$$

One common use of the distributive principle is to remove a common factor or divisor from two terms in a sum or difference.

Example 3   The Distributive Principle

Use distributive principles to replace the sum or difference by an equivalent product or quotient. Do not compute and do not use the commutative principle.

(a) $(6)(3) + (4)(3)$ (b) $\dfrac{x}{5} - \dfrac{15}{5}$

(a) There is a common factor of 3 in each term. We can bring this common factor out and get

$$(6+4)(3).$$

We want to write the multiplying by 3 on the right-hand side, just as it is in the original expression since we are not allowed to use the commutative principle.

(b) There is a common divisor of 5. We have

$$\frac{x}{5} - \frac{15}{5} = \frac{x-15}{5}.$$

EXERCISE 3.7

DEVELOP YOUR SKILL

Use distributive principles to replace each product or quotient by an equivalent sum or difference. Do not compute and do not use the commutative principle.

1. $6(7 + 2)$

2. $(23 - 14)(4)$

3. $\dfrac{27 - 12}{3}$

4. $\dfrac{12 + 18}{a}$

5. $\dfrac{y + x}{7}$

6. $\dfrac{n - 3}{d}$

7. $(a + 4)(5)$

8. $(s)(t - u)$

Use distributive principles to replace each sum or difference by an equivalent product or quotient. Do not compute and do not use the commutative principle.

9. $(9)(7) - (9)(3)$

10. $\dfrac{10}{2} + \dfrac{6}{2}$

11. $8y - 3y$

12. $5x + 5y$

13. $\dfrac{x}{3} - \dfrac{h}{3}$

14. $ab - ac$

15. $6a - 6b$

16. $5n + 7n$

MAINTAIN YOUR SKILL

Use the definition of subtraction to change each subtraction equation to the corresponding addition equation and each addition equation to the corresponding subtraction equation. [1.5]

17. $35 - 17 = 18$

18. $x + 5 = 22$

19. $13 + n = y$

20. $31 - a = b$

Solve each equation using an invariant principle. [1.8, 1.9]

21. $364 + 295 = x + 275$

22. $471 - x = 571 - 388$

Indicate for what counting numbers each statement is true. When possible, list the four smallest values of x. [3.3]

23. 9 divides x

24. x divides 9

3.8 – INVARIANT PRINCIPLE FOR PRODUCTS

OBJECTIVE

1. Use the invariant principle for products.

The invariant principle for products is similar to the invariant principle for sums in Lesson 1.8: if one part gets bigger, the other part must get smaller. For example, we may model the product $4 \cdot 5$ by arranging markers in 4 rows and 5 columns. We see that there are $4 \cdot 5 = 20$ markers in all.

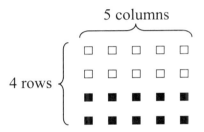

If the top 2 rows are moved to the right and down, then the number of rows is divided by 2 and the number of columns is multiplied by 2. The total number of markers, however, remains the same.

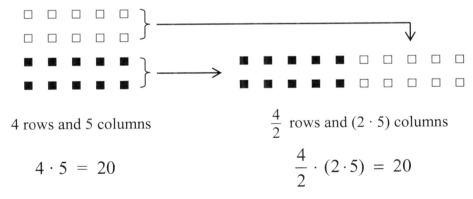

4 rows and 5 columns $\dfrac{4}{2}$ rows and $(2 \cdot 5)$ columns

$4 \cdot 5 = 20$ $\dfrac{4}{2} \cdot (2 \cdot 5) = 20$

This shows the invariant principle for products: if one factor in a product is contracted (divided) by a number and the other factor is expanded (multiplied) by the same number, then the product remains the same.

Invariant Principle for Products

Let x, y and z be counting numbers.

If z divides x, then $xy = \left(\dfrac{x}{z}\right)(yz)$.

If z divides y, then $xy = (xz)\left(\dfrac{y}{z}\right)$.

Example 1   Invariant Principle for Products

If $35 \cdot 3 = x \cdot 15$, what is x?

Let's look at this problem like we did Example 2 in Lesson 1.8. We want to see how the left side of the equation compares to the right side.

$$\overset{\times 5}{\overbrace{}} $$
$$35 \cdot 3 = x \cdot 15$$
$$\underset{?}{\underbrace{}}$$

We note that 3 can be multiplied by 5 to get 15. To keep the product the same, if one factor is multiplied by 5, then the other factor must be divided by 5. That is,

$$x = 35 \div 5 = 7.$$

Example 2   Invariant Principle for Products

If $41 \cdot x = 82 \cdot 24$, what is x?

This time the variable is on the left side, so we look at the change from right to left.

$$41 \cdot x = 82 \cdot 24$$

We note that 82 can be divided by 2 to get 41. To keep the product the same, if one factor is divided by 2, then the other factor must be multiplied by 2. That is,

$$x = 24 \cdot 2 = 48.$$

EXERCISE 3.8

DEVELOP YOUR SKILL

Solve each equation by using the invariant principle for products.

1. $6 \cdot 14 = 12 \cdot x$
2. $9 \cdot 18 = x \cdot 6$
3. $7 \cdot x = 14 \cdot 16$
4. $x \cdot 16 = 12 \cdot 8$
5. $8 \cdot 24 = x \cdot 6$
6. $15 \cdot x = 5 \cdot 21$
7. $x \cdot 25 = 15 \cdot 75$
8. $12 \cdot 15 = 36 \cdot x$

MAINTAIN YOUR SKILL

Each expression is an operand followed by two operators. Do not compute and do not change the operand, but use parentheses to write an equivalent expression with the composite operator. When possible, give an alternate form. [2.3, 2.5]

9. $24 - 9 - 3$
10. $12 + 14 - 18$
11. $15 + 3 - 17$
12. $23 - 16 + 10$
13. $21 - 11 + 15$
14. $12 + 4 + 16$
15. $6 + 9 - 5$
16. $35 - 18 - 4$

Name each form (sum, difference, product, or quotient) and then compute the basic numeral. [3.5]

17. $25 - 10 \cdot 2$

18. $\dfrac{6 \cdot 5}{3}$

19. $3\left(\dfrac{12 - 8}{2}\right)$

20. $(8 - 3)7$

21. $42 \div 7 - 4$

22. $2 \cdot 8 + 3$

23. $\dfrac{18 - 3 \cdot 2}{3}$

24. $9 + 15 \div 3$

3.9 – INVARIANT PRINCIPLE FOR QUOTIENTS

OBJECTIVE

1. Use the invariant principle for quotients.

The invariant principle for quotients is similar to the invariant principle for differences in Lesson 1.9: both parts change in the same way. For example, when 12 markers are divided into 6 equal parts, each part contains 2:

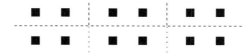

If we just look at the top part of the diagram, we have divided the number of markers by 2 and the number of parts by 2. We see that each part still contains 2:

This shows that $\dfrac{12}{6}$ is the same as $\dfrac{12 \div 2}{6 \div 2}$ or $\dfrac{6}{3}$, since both are equal to 2.

Similarly, if we start with 6 and divide it into 3 equal parts, each part will contain 2:

If we double this diagram we see that when $6 \cdot 2$ is divided into $3 \cdot 2$ equal parts, each part still contains 2:

This shows that $\dfrac{6}{3}$ is the same as $\dfrac{6 \cdot 2}{3 \cdot 2}$ or $\dfrac{12}{6}$.

Example 1   Invariant Principle for Quotients

(a) *Divide both the dividend and the divisor of $\dfrac{12}{6}$ by 3 to get an equivalent quotient.*

(b) *Multiply both the dividend and the divisor of $\dfrac{12}{6}$ by 2 to get an equivalent quotient.*

(a) $\dfrac{12}{6} = \dfrac{12 \div 3}{6 \div 3} = \dfrac{4}{2}$ (b) $\dfrac{12}{6} = \dfrac{12 \cdot 2}{6 \cdot 2} = \dfrac{24}{12}$

Note that the value of all these quotients is 2:

$$\frac{12}{6} = \frac{4}{2} = \frac{24}{12} = 2$$

These examples illustrate the invariant principle for quotients: if both the dividend and the divisor in a quotient are divided by the same number, the quotient remains the same. Likewise, both the dividend and the divisor may be multiplied by the same number and the quotient remains the same.

Invariant Principle for Quotients

> Let x, y and z be counting numbers with $z \neq 0$. If y divides x and z divides both x and y, then $\dfrac{x}{y} = \dfrac{x \div z}{y \div z}$.
>
> If y divides x, then $\dfrac{x}{y} = \dfrac{x \cdot z}{y \cdot z}$.

Example 2   **Invariant Principle for Quotients**

Solve the equation $\dfrac{12}{3} = \dfrac{x}{15}$.

When we compare the divisors, we see that 3 has been multiplied by 5 to get 15.

$$\overset{?}{\underset{\times 5}{\dfrac{12}{3} = \dfrac{x}{15}}}$$

In order to keep the quotient the same, 12 must be multiplied by 5 to get x. That is,

$$x = 12 \cdot 5 = 60.$$

Example 3   Invariant Principle for Quotients

Solve the equation $\dfrac{x}{8} = \dfrac{120}{24}$.

We need to look from right to left since the variable x is on the left.

$$\frac{x}{8} = \frac{120}{24}$$

$\div\, 3$

In moving from 24 to 8 in the divisors, we have divided by 3. To keep the same value for both quotients, in moving from 120 to x we must also divide by 3. That is,

$$x = 120 \div 3 = 40.$$

Study Tip:

Products behave like sums: if one term gets smaller, the other term must get larger to maintain the same product.

Quotients behave like differences: both terms change in the same way.

Example 4   Use an Invariant Principle

Solve the equation $15 \cdot 7 = 3 \cdot x$.

This time we use the invariant principle for products, as in Lesson 3.8. We see that 15 is divided by 5 to get 3.

$\div\, 5$

$$15 \cdot 7 = 3 \cdot x$$

?

To keep the product the same, the 7 must be multiplied by 5 to get x. That is,

$$x = 7 \cdot 5 = 35.$$

Exercise 3.8

Develop Your Skill

1. Multiply both the dividend and the divisor of $\dfrac{24}{8}$ by 2 to get an equivalent quotient.

2. Divide both the dividend and the divisor of $\dfrac{24}{8}$ by 4 to get an equivalent quotient.

Solve each equation by using an invariant principle.

3. $\dfrac{35}{7} = \dfrac{x}{14}$

4. $\dfrac{63}{9} = \dfrac{21}{x}$

5. $40 \div x = 8 \div 2$

6. $\dfrac{x}{12} = \dfrac{90}{6}$

7. $24 \cdot x = 12 \cdot 22$

8. $45 \cdot 15 = 9 \cdot x$

9. $\dfrac{70}{14} = \dfrac{210}{x}$

10. $\dfrac{40}{x} = \dfrac{80}{16}$

11. $24 \cdot 5 = x \cdot 20$

12. $126 \div 6 = x \div 3$

Maintain Your Skill

Use the composition of operators to write an equivalent single operator. [2.5]

13. $+ m - n$, where $n \geq m$

14. $- n + m$, where $m \geq n$

Use distributive principles to replace each product or quotient by an equivalent sum or difference and to replace each sum or difference by an equivalent product or quotient. Do not compute and do not use the commutative principle. [3.7]

15. $7(8 + 3)$

16. $(8)(3) - (4)(3)$

17. $\dfrac{32}{4} - \dfrac{24}{4}$

18. $\dfrac{18 + 24}{3}$

19. $3x + 3y$

20. $7y - 4y$

21. $\dfrac{a-c}{bd}$

22. $\dfrac{28}{2x} - \dfrac{n}{2x}$

23. $5w + nw$

24. $(3b)(x-5)$

CHAPTER 3 REVIEW

1. Construct a graphical model for the expansion operator $7(\;\;)$. [3.1]

2. Construct a graphical model for the contraction operator $\dfrac{(\;\;)}{6}$. [3.2]

For each equation, write the corresponding equation that is matched to it by the definition of division. [3.2]

3. $\dfrac{56}{7} = 8$

4. $3x = y$

5. $n \div c = 30$

Indicate for what counting numbers each statement is true. When possible, list the four smallest values of x. [3.3]

6. 6 divides x

7. x divides 10

8. 0 divides x

9. x divides 0

Indicate for what counting numbers each quotient is defined. When possible, list the four smallest values of x. [3.3]

10. $\dfrac{20}{x}$

11. $\dfrac{x}{20}$

Determine whether each number is divisible by 2, 3, 5, or 10. [3.4]

12. 1250

13. 1784

14. 8745

Name each form and then compute the basic numeral. [3.5]

15. $4(7-2)$

16. $3 \cdot 5 + \dfrac{18}{3}$

17. $2 + 3 \cdot 4$

18. $6 \cdot 2 - 7$

19. $10 \div 2 - 9 \div 3$

20. $15 - (2 \cdot 4)$

21. $4\left(\dfrac{35}{5}\right)$ **22.** $2 \cdot 20 \div 4$

Write all other possible equivalent forms that result from using the commutative principles for addition and multiplication. If there are none, write "none." Do not compute. [3.6]

23. $\dfrac{36}{4+5}$ **24.** $21 - nx$ **25.** $7(8-5)$

26. $\dfrac{x-6}{3}$ **27.** $\dfrac{x}{y}+3$ **28.** $3\left(\dfrac{x}{4}\right)$

Use distributive principles to replace each product or quotient by an equivalent sum or difference, and to replace each sum or difference by an equivalent product or quotient. Do <u>not</u> compute and do not use the commutative principle. [3.7]

29. $5x + 5y$ **30.** $\dfrac{n}{2} - \dfrac{12}{2}$ **31.** $7(x-b)$

32. $\dfrac{x+12}{4}$ **33.** $(x+5)y$ **34.** $9n - 3n$

Solve each equation by using an invariant principle. [3.8, 3.9]

35. $x \cdot 3 = 244 \cdot 6$ **36.** $\dfrac{x}{15} = \dfrac{105}{5}$

I thought we weren't going to use fractions. What about $\dfrac{12}{2}$?

That's not a fraction. It's a quotient. $\dfrac{12}{2}$ is just another name for the natural number 6.

Chapter 4 – Expansions and Contractions

4.1 – COMPOSITION OF TWO EXPANSIONS OR TWO CONTRACTIONS

OBJECTIVE

1. Combine two expansions or two contractions and model their actions by comparing line segments.

The compositions of expansions and contractions follow a set of rules that are very similar to the rules for increases and decreases. (See Lessons 2.2 and 2.4.) In this lesson we look at the composition of two expansions or two contractions. In Lesson 4.2 we consider how they interact with each other.

Example 1   **The Composition of Two Expansions**

Three friends are comparing how many baseball cards they have. Doug has 2 cards, Rick has 8 cards and Matt has 24.

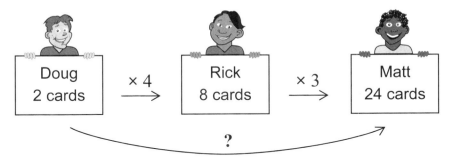

To compare the number of Doug's cards to Rick's, we multiply by 4. To compare the number of Rick's cards to Matt's, we multiply by 3. How can we compare the number of Doug's cards to Matt's?

We see that the number of Doug's cards must be multiplied by 12 to get the number of Matt's cards. Where did the 12 come from? It came from multiplying 4 times 3.

Now let's model the expansions from Example 1 using line segments.

<u>Case 1</u>   The composition of two expansions.

Here is the composition of multiplying by 4 and then multiplying by 3.

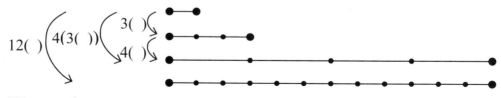

We see that multiplying by 4 and then multiplying by 3 is the same as multiplying by their product 12.

$$3(4(\ )) = (3 \cdot 4)(\ )$$

If we multiply first by 3 and then by 4, we obtain the same result.

We see that

$$3(4(\ )) = (3 \cdot 4)(\ ) = (4 \cdot 3)(\ ) = 4(3(\ )).$$

Here is the general rule:

<u>Case 1</u>   An expansion by *a* and an expansion by *b*.

The composite of two expansions (multiplying by *a* and multiplying by *b*) is equivalent to the expansion resulting from multiplying by the product *ab* (or multiplying by *ba*). The two expansions can be in either order.

$$b(a(\ )) = ab(\ ) = ba(\ ) = a(b(\ ))$$

Study Tip:

When using parentheses, we start on the inside and work out. That's why 3(4( )) means multiply first by 4 and then by 3.

Example 2   The Composition of Two Contractions

Let's compare the number of baseball cards from Example 1 in a different order, and suppose that Doug now has 4 cards.

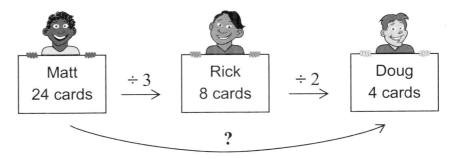

To compare the number of Matt's cards to Rick's, we divide by 3. *To compare the number of Rick's cards to Doug's, we divide by* 2. *How can we compare the number of Matt's cards to Doug's?*

We see that the number of Matt's cards must be divided by 6 to get the number of Doug's cards. Where did the 6 come from? It came from multiplying 3 times 2.

Case 2   The composition of two contractions.

Here is the composition of dividing by 3 and then dividing by 2.

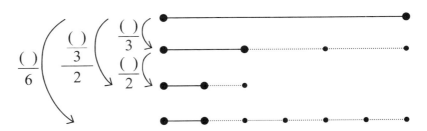

We see that dividing by 3 and then dividing by 2 is the same as dividing by their product 6.

$$\frac{\dfrac{(\ )}{3}}{2} = \frac{(\ )}{3 \cdot 2}$$

Also, if we divide first by 2 and then by 3, the result is the same.

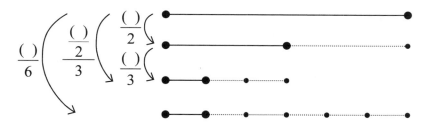

We see that

$$\frac{\dfrac{(\ )}{3}}{2} = \frac{(\ )}{3 \cdot 2} = \frac{(\ )}{2 \cdot 3} = \frac{\dfrac{(\ )}{2}}{3}.$$

Here is the general rule:

Case 2   A contraction by a and a contraction by b.

The composite of two contractions (dividing by a and dividing by b) is equivalent to the contraction resulting from dividing by the product ab (or dividing by ba). The two contractions can be in either order.

$$\frac{\dfrac{(\ )}{a}}{b} = \frac{(\ )}{ab} = \frac{(\ )}{ba} = \frac{\dfrac{(\ )}{b}}{a}$$

Example 3   **Describe Operators**

Write a sentence that describes the action of the two operators.

(a) $5(3(\ ))$ (b) $\dfrac{\dfrac{(\ )}{6}}{2}$

(a) Expand by 3 and then expand by 5.
(b) Contract by 6 and then contract by 2.

Example 4   **Equivalent Operators**

Write an equivalent operator in basic form.

(a) $5(3(\ ))$ (b) $\dfrac{\dfrac{(\ )}{6}}{2}$ (c) $\dfrac{(\ )}{\dfrac{6}{2}}$

(a) The two expansions work together to give a larger expansion.

$$5(3(\ )) = (5 \cdot 3)(\ ) = 15(\ )$$

(b) The two contractions work together to give a larger contraction.

$$\frac{\dfrac{(\ )}{6}}{2} = \frac{(\ )}{6 \cdot 2} = \frac{(\ )}{12}$$

(c) This looks similar to part (b), but there is an important difference between them. In part (b), the longer division bar is at the bottom. This means we first divide by 6 and then divide by 2. But in part (c), the longer division bar is at the top. This groups the bottom part together:

$$\frac{(\ )}{\frac{6}{2}} = \frac{(\ )}{\left(\frac{6}{2}\right)} = \frac{(\ )}{3}$$

We compute the 6 divided by 2, and get a contraction by 3.

We may summarize the composition of two expansions or two contractions as follows:

Composition of Two Expansions or Two Contractions	The composition of two expansions or two contractions work together to give a stronger composite change of the same kind. Multiply their magnitudes.

EXERCISE 4.1

DEVELOP YOUR SKILL

Write a sentence that describes the action of the operators. Then construct a graphical model for their composition.

1. $4(2(\ ))$

2. $\dfrac{\dfrac{(\ )}{2}}{4}$

Write an equivalent operator in basic form.

3. $3(8(\ ))$

4. $9(2(\ ))$

5. $\dfrac{\dfrac{(\ )}{8}}{2}$

6. $\dfrac{(\ )}{\dfrac{8}{2}}$

7. $\dfrac{(\ )}{\dfrac{12}{3}}$

8. $\dfrac{\dfrac{(\ )}{12}}{3}$

9. $m(n(\ ))$

10. $\dfrac{\dfrac{(\ )}{m}}{n}$

MAINTAIN YOUR SKILL

11. The coach had fifty-six players. He wanted to make eight equal teams. How many players should he put on each team? [3.2]

For each equation, write the corresponding equation that is matched to it by the definition of subtraction or the definition of division. [1.5, 3.2]

12. $26 - 8 = 18$

14. $7 + 8 = 15$

14. $\dfrac{28}{4} = 7$

15. $(6)(5) = 30$

Name each form (sum, difference, product, or quotient) and then compute the basic numeral. [3.5]

16. $7 + \dfrac{12}{3}$

17. $5 \cdot 7 - 6$

18. $\dfrac{24}{3 + 5}$

19. $3 + 4 \cdot 5$

20. $5\left(\dfrac{15 + 6}{3}\right)$

21. $\dfrac{6 \cdot 5}{5 - 2}$

Use distributive principles to replace each product or quotient by an equivalent sum or difference, and to replace each sum or difference by an equivalent product or quotient. Do <u>not</u> compute and do not use the commutative principle. [3.7]

22. $5(x + 3)$

23. $\dfrac{15}{3} + \dfrac{h}{3}$

24. $7x - 7b$

25. $6a - 2a$

4.2 – COMPOSITION OF AN EXPANSION AND A CONTRACTION

OBJECTIVE

1. Combine expansions and contractions and model their actions by comparing line segments.

The last lesson showed how to combine two expansions or two contractions: the result is a larger change of the same kind. When combining an expansion with a contraction, the two changes work against each other, so the composite will be smaller than either of the individual changes.

Example 1   **The Composition of an Expansion and a Contraction**

Let's return to Example 2 in Lesson 4.1 and suppose that Rick now has 12 cards.

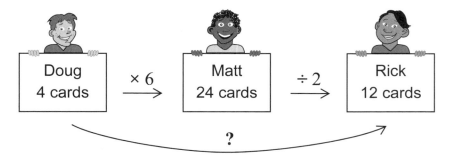

To compare the number of Doug's cards to Matt's, we multiply by 6. To compare the number of Matt's cards to Rick's, we divide by 2. How can we compare the number of Doug's cards to Rick's?

We see that the number of Doug's cards must be multiplied by 3 to get the number of Rick's cards. Where did the 3 come from? Let's look at the model with line segments and see.

<u>Case 3</u>   The composition of an expansion and a contraction, when the expansion is larger.

Here is the composition of multiplying by 6 and then dividing by 2. Since the expansion is stronger than the contraction, we expect the combined change to be an expansion.

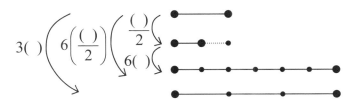

We see that multiplying by 6 and then dividing by 2 is the same as multiplying by 3. The magnitude 3 of the composition comes from taking the larger magnitude 6 and dividing by the smaller magnitude 2.

$$\frac{6(\ )}{2} = \frac{6}{2}(\ )$$

Also, if we first divide by 2 and then multiply by 6, the result is the same.

We see that

$$\frac{6(\ )}{2} = \frac{6}{2}(\ ) = 6\left(\frac{(\ )}{2}\right).$$

Study Question:

What is the difference between $\frac{6(\ )}{2}$ and $6\left(\frac{(\ )}{2}\right)$?

Answer:

$\frac{6(\ )}{2}$ tell us to multiply by 6 and then divide by 2.

$6\left(\frac{(\ )}{2}\right)$ says to divide by 2 first and then multiply by 6.

Here is the general rule:

<u>Case 3</u>   An expansion by a and a contraction by b, where b divides a.

The composite of an expansion (multiplying by a) and a contraction (dividing by b), where b divides a, is equivalent to the expansion resulting from multiplying by $\dfrac{a}{b}$. The two changes can be in either order.

$$\frac{a(\ )}{b} = \frac{a}{b}(\ ) = a\left(\frac{(\ )}{b}\right)$$

Example 2   **The Composition of an Expansion and a Contraction**

Let's look one more time at the three friends comparing how many baseball cards they have. This time, Rick has 8 cards.

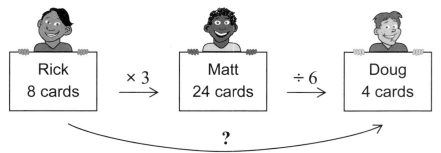

To compare the number of Rick's cards to Matt's, we multiply by 3. To compare the number of Matt's cards to Doug's, we divide by 6. How can we compare the number of Rick's cards to Doug's?

We see that the number of Rick's cards must be divided by 2 to get the number of Rick's cards. Where did the 2 come from? Let's look at the model with line segments and see.

<u>Case 4</u>   The composition of an expansion and a contraction, when the contraction is larger.

Here is the composition of multiplying by 3 and then dividing by 6. Since the contraction is stronger than the expansion, we expect the combined change to be a contraction.

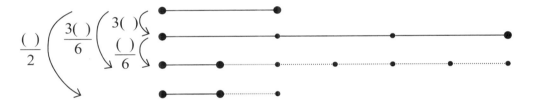

We see that multiplying by 3 and then dividing by 6 is the same as dividing by 2. The magnitude 2 of the composition comes from taking the larger magnitude 6 and dividing by the smaller magnitude 3.

$$\frac{3(\ )}{6} = \frac{(\ )}{\frac{6}{3}}$$

Also, if we first divide by 6 and then multiply by 3, the result is the same.

We see that

$$\frac{3(\ )}{6} = \frac{(\ )}{\frac{6}{3}} = 3\left(\frac{(\ )}{6}\right).$$

Here is the general rule:

<u>Case 4</u>   An expansion by a and a contraction by b, where a divides b.

The composite of an expansion (multiplying by a) and a contraction (dividing by b), where a divides b, is equivalent to the contraction resulting from dividing by $\frac{b}{a}$. The two changes can be in either order.

$$\frac{a(\ )}{b} = \frac{(\ )}{\frac{b}{a}} = a\left(\frac{(\ )}{b}\right)$$

We may summarize the results from the last two lessons as follows:

Composition of Expansions and Contractions

---

1. Like changes help each other to give a stronger composite change of the same kind. Multiply their magnitudes.

2. Expansions and contractions work against each other. The stronger change will dominate the weaker change and determine the kind of composite change. Divide the larger magnitude by the smaller.

Example 3   Find a Composite Operator

Write a sentence that describes the action of the two operators. Then compute the basic form for the composite operator, when possible. If the basic form is not yet defined, write "not defined."

(a) $4\left(\dfrac{(\ )}{20}\right)$ (b) $\dfrac{\frac{(\ )}{7}}{4}$ (c) $\dfrac{20(\ )}{5}$ (d) $\dfrac{4(\ )}{3}$

(a) Contract by 20 and then expand by 4. The contraction $\dfrac{(\ )}{20}$ and the expansion $4(\ )$ are working against each other. The contraction is stronger, so the composite change will be a contraction. We divide the magnitudes.

$$4\left(\frac{(\ )}{20}\right) = \frac{\frac{(\ )}{20}}{4} = \frac{(\ )}{5}$$

(b) Contract by 7 and then contract by 4. The two contractions $\dfrac{(\ )}{7}$ and $\dfrac{(\ )}{4}$ work together to give a larger contraction. Multiply their magnitudes.

$$\frac{\frac{(\ )}{7}}{4} = \frac{(\ )}{7 \cdot 4} = \frac{(\ )}{28}.$$

(c) Expand by 20 and then contract by 5. The expansion $20(\ )$ and the contraction $\dfrac{(\ )}{5}$ are working against each

Caution:

The answer to part (c) is an expansion 4( ), not just the number 4.

other. This time the expansion is stronger, so the composite change will be an expansion.   We divide the magnitudes.

$$\frac{20(\ )}{5} = \frac{20}{5}(\ ) = 4(\ ).$$

(d) Expand by 4 and then contract by 3. The composition $\frac{4(\ )}{3}$ is not defined yet, since 3 does not divide 4 and 4 does not divide 3.

EXERCISE 4.2

DEVELOP YOUR SKILL

Construct a graphical model for the composition of the operators.

1. $4\left(\dfrac{(\ )}{2}\right)$

2. $\dfrac{2(\ )}{4}$

Write a sentence that describes the action of the two operators. Then compute the basic form of the composite operator, when possible. If this basic form is not yet defined, write "not defined."

3. $\dfrac{12(\ )}{4}$

4. $5(2(\ ))$

5. $\dfrac{3(\ )}{8}$

6. $4\left(\dfrac{(\ )}{8}\right)$

7. $\dfrac{\frac{(\ )}{12}}{4}$

8. $\dfrac{\frac{(\ )}{4}}{12}$

9. $15\left(\dfrac{(\ )}{3}\right)$

10. $\dfrac{2(\ )}{16}$

11. $5\left(\dfrac{(\ )}{3}\right)$

12. $\dfrac{\dfrac{(\ )}{5}}{3}$

13. $8\left(\dfrac{(\ )}{2}\right)$

14. $\dfrac{20(\ )}{5}$

MAINTAIN YOUR SKILL

15. As the storm was approaching, the tide rose 2 feet. Then it rose 6 feet more, and then an additional 4 feet. After the storm passed, it fell 7 feet. Construct an operator equation to model the changes in the tide over this period of time. Put a sequence of operators on the left side of the equation and the composite operator on the right side. [2.4]

16. (a) What is the smallest x such that x divides both 18 and 24? [3.3]

(b) What is the largest x such that x divides both 18 and 24?

(c) What is the smallest nonzero x such that 18 and 24 both divide x?

Determine whether each number is divisible by 2, 3, 5 and/or 10. [3.4]

17. 345

18. 456

19. 468

20. 7,890

Solve each equation by using an invariant principle. [3.8, 3.9]

21. $33 \cdot 12 = 11 \cdot x$

22. $28 \cdot x = 14 \cdot 22$

23. $\dfrac{60}{x} = \dfrac{20}{4}$

24. $\dfrac{36}{12} = \dfrac{x}{4}$

4.3 – THE COMMUTATIVE PRINCIPLE

OBJECTIVE

1. Use the commutative principle for expansions and contractions.

In Lessons 4.1 and 4.2 we saw that the order of applying expansions and contractions can be interchanged without changing the result. This is another example of the commutative principle.

Example 1   Describe Operators

Write a sentence that describes the action of the two operators.

(a) $5(6(\ ))$ (b) $\dfrac{3(\ )}{6}$ (c) $10\left(\dfrac{(\ )}{2}\right)$

(a) Expand by 6 and then expand by 5.

(b) Expand by 3 and then contract by 6.

(c) Contract by 2 and then expand by 10.

Example 2   Commute Operators

Commute the order of the operators in Example 1. Then compute the basic form.

(a) $5(6(\ ))$ (b) $\dfrac{3(\ )}{6}$ (c) $10\left(\dfrac{(\ )}{2}\right)$

(a) Instead of expanding by 6 and then expanding by 5, we want to expand by 5 first, then expand by 6. That puts the expansion by 5 on the "inside:" $6(5(\ ))$. In both cases the two expansions are working together to give an expansion of their product: 30( ).

(b) Instead of expanding by 3 and then contracting by 6, we want to contract by 6 first, then expand by 3. So we start with $\dfrac{(\,)}{6}$ and then apply 3( ) to get $3\left(\dfrac{(\,)}{6}\right)$. In both cases the expansion and the contraction are working against each other with the contraction larger. We get a contraction of their quotient: $\dfrac{(\,)}{2}$.

(c) Instead of contracting by 2 and then expanding by 10, we want to expand by 10 first, then contract by 2. So we start with 10( ) and then apply $\dfrac{(\,)}{2}$ to get $\dfrac{10(\,)}{2}$. This time the expansion is larger so the combined change is an expansion: 5( ).

EXERCISE 4.3

DEVELOP YOUR SKILL

Write a sentence that describes the action of the two operators.

1. $3\left(\dfrac{(\,)}{15}\right)$

2. $3(8(\,))$

3. $20\left(\dfrac{(\,)}{4}\right)$

4. $\dfrac{\dfrac{(\,)}{10}}{2}$

5. $\dfrac{6(\,)}{18}$

6. $\dfrac{45(\,)}{9}$

Commute the order of the two operators.

7. $3\left(\dfrac{(\ )}{15}\right)$

8. $3(8(\ ))$

9. $20\left(\dfrac{(\ )}{4}\right)$

10. $\dfrac{\dfrac{(\ )}{10}}{2}$

11. $\dfrac{6(\ )}{18}$

12. $\dfrac{45(\ )}{9}$

Compute the basic form of each composite operator.

13. $3\left(\dfrac{(\ )}{15}\right)$

14. $3(8(\ ))$

15. $20\left(\dfrac{(\ )}{4}\right)$

16. $\dfrac{\dfrac{(\ )}{10}}{2}$

17. $\dfrac{6(\ )}{18}$

18. $\dfrac{45(\ )}{9}$

MAINTAIN YOUR SKILL

19. Kim bought 6 colored pencils for 28 cents each. If she paid $2, how much did she get back in change? [3.1]

Each expression is an operand followed by two operators. Do not change the operand, but use parentheses to write an equivalent expression with the composite operator. Assume that $b \geq c$ and that all differences are defined. When possible, give an alternate form. [2.3, 2.5]

20. $25 - c + b$

21. $25 + b + c$

22. $25 - c - b$

23. $25 + b - c$

24. $25 + c - b$

25. $25 - b + c$

4.4 – BUILDING COMPOUND EXPRESSIONS

OBJECTIVE

1. Build compound expressions one step at a time.

    Sometimes algebraic expressions contain several different operations. In analyzing a compound expression it is helpful to think of it as having been built up one step at a time by a sequence of operators. The operators we have seen so far are

$$+ n, \quad n +, \quad - n, \quad n(\ ), \quad (\ )n, \quad \text{and} \quad \frac{(\ )}{n} \text{ when } n \neq 0.$$

We begin by applying a sequence of operators to an operand to see how a compound expression may be constructed.

Example 1   **Build a Compound Expression**

An operand is given below, followed by operators to be used in the order stated. Build this form, but do not compute and do not use the commutative principle. Then name the form that has been built.

$$15; \quad - 3, \quad \frac{(\ )}{2}, \quad + 5, \quad 2(\ )$$

Start with 15. Then subtract 3: $15 - 3$

Then divide by 2: $\dfrac{15 - 3}{2}$

Then add 5 (on the right): $\dfrac{15 - 3}{2} + 5$

Then multiply (on the left) by 2: $2\left(\dfrac{15 - 3}{2} + 5\right)$

The last operation is multiplication, so this is a product.

Remember:

The **name** of a form comes from the **last** operation used in building it.

Study Tip:

At each step the operator is applied to the whole expression that has already been built. This is why we don't get

$$\frac{1+(5)(3)-2}{4}$$

or $\frac{1+(5)(3)}{4-2}$.

Example 2   Build a Compound Expression

An operand is given below, followed by operators to be used in the order stated. Build this form, but do not compute and do not use the commutative principle. Then name the form that has been built.

$$5; \ (\ )3, \ 1+, \ \frac{(\ )}{4}, \ -2$$

Start with 5. Then multiply (on the right) by 3: $(5)(3)$

Then add 1 (on the left): $1 + (5)(3)$

Then divide by 4: $\dfrac{1+(5)(3)}{4}$

Then subtract 2: $\dfrac{1+(5)(3)}{4} - 2$

The last operation is subtraction, so this is a difference.

EXERCISE 4.4

DEVELOP YOUR SKILL

In Exercises 1 – 10, an operand number is given, followed by operators to be used in the order stated. Build this form, but do not compute and do not use the commutative principle. Then name the form that has been built.

1. $17; \ +1, \ \dfrac{(\ )}{3}, \ -2$
2. $5; \ (\ )3, \ +1, \ \dfrac{(\ )}{2}$

3. $8; \ +4, \ 3(\ ), \ \dfrac{(\ )}{2}$
4. $15; \ \dfrac{(\ )}{3}, \ +1, \ 4(\ )$

5. $12; \ -2, \ \dfrac{(\ )}{5}, \ +6$
6. $7; \ 5+, \ 3(\ ), \ -5$

7. 18; $\dfrac{(\ )}{2}$, -4, $(\ )3$ **8.** 2; $+10$, $\dfrac{(\ )}{3}$, $5(\ )$

9. 15; -3, $\dfrac{(\ )}{4}$, $+7$ **10.** 7; $3+$, $2(\ )$, $\dfrac{(\ )}{5}$

Maintain Your Skill

Use an invariant principle to write an equivalent sum or difference that is easier to compute. Then compute. [1.8, 1.9]

11. $297 + 256$ **12.** $523 - 398$

See each expression as having one operand and one operator. List all the ways to name the operand and the operator. [2.1, 3.2]

13. $19 - 5$ **14.** $5 + 7$ **15.** $(2)(8)$

16. $\dfrac{18}{3}$ **17.** $(8 - 3)(7)$ **18.** $\dfrac{14 + 4}{6}$

Construct a vector model for the composition of the operators. [2.4]

19. $+6 - 8$ **20.** $-4 + 7$

Write the composition of the two given operators as an equivalent single operator.
[2.4 and 4.2]

21. $+x - y$, if $x \geq y$ **22.** $+x - y$, if $y \geq x$

23. $\dfrac{x(\ )}{y}$, if y divides x **24.** $\dfrac{x(\ )}{y}$, if x divides y

4.5 – Analyzing Compound Expressions

OBJECTIVE

1. Analyze a compound expression as a sequence of operators.

In practice, we often want to work backwards from what we did in Lesson 4.4. That is, we are given a compound expression and we want to see how to build it by applying a sequence of operators to a single operand. To do this, we begin as if computing it, remembering the order of operations.

Remember the order of operations:

Multiplication and division are computed before addition and subtraction, unless directed otherwise by parentheses.

Study Tip:

In (b), we might think of starting with m as the operand. But then the operator would be $+\frac{x}{y}$, which

is not a basic operator since it is made up of two parts.

Example 1   Analyze an Expression

Analyze each expression by stating a basic operand and then listing a sequence of basic operators that can be joined one at a time to produce it.

(a) $5 \cdot 7 - 4$ (b) $m + \dfrac{x}{y}$ (c) $u\left(\dfrac{w}{x} - y\right) + z$

(a) We do multiplication before subtraction, so we start with $5 \cdot 7$. There are two choices: use 5 as the operand and apply the operators $(\ )7$ and -4, or use 7 as the operand and apply the operators $5(\ )$ and -4. So we have two answers:

$$5;\ (\ )7,\ -4 \quad \text{or} \quad 7;\ 5(\ ),\ -4$$

Note that $(\ )7$ multiplies by 7 on the right and $5(\ )$ multiplies by 5 on the left.

(b) Division is done before addition, so we must start with the operand x and apply the operators $\dfrac{(\ )}{y}$ and $m+$, in that order. Note that the m is added on the left side:

$$x;\ \dfrac{(\ )}{y},\ m+$$

(c) We start inside the parentheses and division is done before subtraction. So the initial operand must have been w. The operators are $\dfrac{(\ )}{x}$, $-y$, $u(\ )$, and $+z$, in that order.

Note that we multiply by u on the left and add z on the right.

$$w;\ \dfrac{(\ )}{x},\ -y,\ u(\ ),\ +z$$

EXERCISE 4.5

DEVELOP YOUR SKILL

Analyze each expression by stating a basic operand and then listing a sequence of basic operators that can be joined one at a time to produce the given expression. If more than one operand and sequence of operators can be used, give the alternate possibilities.

1. $12 - 5$

2. $3 \cdot 8$

3. $4(9 - 3)$

4. $\dfrac{18}{6} + 5$

5. $5 \cdot 6 - 7$

6. $\dfrac{4 + 5}{3}$

7. $6 + \dfrac{10}{2}$

8. $(8 - 4)(3)$

9. $\dfrac{a}{b} + c$

10. $c(d - e)$

11. $\dfrac{f + g}{h}$

12. $\dfrac{x}{y} - z$

MAINTAIN YOUR SKILL

Each expression is an operand followed by two operators. Do not compute and do not change the operand, but use parentheses to write an equivalent expression with the composite operator. When possible, give an alternate form. [2.3, 2.5]

13. $10 + 4 - 9$

14. $15 - 7 + 2$

15. $21 - 8 - 7$

16. $19 - 3 + 5$

17. $11 + 8 - 3$

18. $4 + 7 + 9$

Use distributive principles to replace the product by an equivalent difference and to replace the sums by an equivalent quotient or product. Do not compute and do not use the commutative principle. [3.7]

19. $(12 - 3)4$

20. $\dfrac{15}{3} + \dfrac{12}{3}$

21. $3n + 8n$ **22.** $5x + 5y$

Write a sentence that describes the action of the two operators. Then construct a graphical model for the composition of the operators. [4.2]

23. $\dfrac{3(\ )}{6}$ **24.** $\dfrac{6(\ )}{3}$

4.6 – CANCELING OPERATORS

OBJECTIVE

1. Cancel operators.

In Lesson 2.6 we saw that the operators $+n$ and $-n$ are **inverse** to each other. That is, when one is applied immediately after the other, the combined result is no change. We can show this with the operator equations

$$+n - n = +0 \quad \text{and} \quad -n + n = +0.$$

A similar thing happens when an expansion and a contraction by the same amount are applied one after the other. The result is no change. For example, multiplying by 2 and then dividing by 2 leaves the operand unchanged. This is the same as multiplying by 1:

$$\frac{2(\ )}{2} = 1(\ )$$

In general, the operators $n(\ )$ and $\dfrac{(\ )}{n}$ are **inverse** to each other:

$$\frac{n(\ )}{n} = 1(\ ) \quad \text{and} \quad n\!\left(\frac{(\ )}{n}\right) = 1(\ )$$

an expansion by n followed by a contraction by n	a contraction by n followed by an expansion by n

Example 1   **Inverse Operators**

Write the inverse of each operator.

(a) $+6$ (b) $4(\ )$ (c) -8 (d) $\dfrac{(\ )}{5}$

(a) The inverse of an increase of 6 is a decrease of 6: -6.

(b) The inverse of an expansion by 4 is a contraction by 4:
$$\frac{(\ )}{4}.$$

(c) The inverse of a decrease of 8 is an increase of 8: $+8$.

(d) The inverse of a contraction by 5 is an expansion by 5:
$$5(\ ).$$

In Example 2 we simplify several expressions by the removal of two inverse operators that are joined in immediate succession. This process is called **canceling** the operators. In order to be inverse to each other, two operators must be members of the same family. That is, increases and decreases can cancel each other if they have the same magnitude and occur one right after the other. Likewise, expansions and contractions can cancel each other if they have the same magnitude and occur one right after the other. But an increase or a decrease can never cancel with a contraction or an expansion.

Study Tip:

To cancel two operators, they must belong to the same family – the add/subtract family or the multiply/divide family.

Example 2   **Canceling Operators**

(a) $7+4-4 = 7$ In general, $x+y-y = x$.

(b) $7-4+4 = 7$ In general, if $y \le x$, then $x-y+y = x$.

(c) $\dfrac{4\cdot 7}{4} = 7$ In general, if $y \neq 0$, then $\dfrac{y(x)}{y} = x$.

(d) $4\left(\dfrac{12}{4}\right) = 12$ In general, if y divides x, then $y\left(\dfrac{x}{y}\right) = x$.

It is operators, but never numbers, that cancel each other. If only the 4's are removed from $\dfrac{4\cdot 7}{4}$, the result is $\dfrac{\cdot 7}{}$, which is nonsense. It is the operators $4\cdot$ and $\dfrac{(\ )}{4}$ that can be canceled to get 7. Dividing by 4 undoes (cancels) the multiplying by 4.

Canceling

Operators

In order to cancel two operators in an expression, the operators must:

(1) Be inverses of each other.

(2) Occur in immediate succession.

If inverse operators such as $+n$ and $-n$ occur in an expression, but they are not used in succession, it is sometimes possible to rearrange the order of operations and then cancel them. This rearranging can be done by using the commutative and associative properties.

Example 3   **Simplify an Expression by Canceling Operators**

Simplify each expression, if possible.

(a) $\dfrac{x \cdot 5 \cdot y}{5}$ (b) $7 + x - 7$ (c) $\dfrac{x+5}{5} - 5$

(a) We have $\dfrac{x \cdot 5 \cdot y}{5} = \dfrac{5 \cdot x \cdot y}{5} = \dfrac{5(x \cdot y)}{5}$ Here we have used both the commutative and the associative properties. In the last expression, we can start with the operand $x \cdot y$.

The operators are $5(\ )$ and $\dfrac{(\ )}{5}$, which are inverse operators used one right after the other. They can be canceled:

$$\dfrac{5(x \cdot y)}{5} = x \cdot y$$

(b) By replacing $7 + x$ with $x + 7$ we obtain

$$7 + x - 7 = x + 7 - 7$$

Now the increase of 7 and the decrease of 7 follow one after the other and they can be canceled:

$$x + 7 - 7 = x.$$

(c) There are three operators in the expression $\dfrac{x+5}{5} - 5$.

The operators $+ 5$ and $\dfrac{(\ )}{5}$ are not inverses. While $+ 5$ and $- 5$ are inverse operators, they are not used in succession. If we were to analyze this expression as in Lesson 4.5, we could start with the operand x. The operators would be

$$+ 5, \quad \dfrac{(\ )}{5}, \quad \text{and} \quad - 5, \quad \text{in that order.}$$

We see that the operator $\dfrac{(\ )}{5}$ is used between $+5$ and -5.

But contractions do not commute with increases and decreases, so there is no way to get the increase of 5 and the decrease of 5 together. This means no cancellation is possible.

EXERCISE 4.6

DEVELOP YOUR SKILL

Write the inverse of each operator.

1. -12
2. $\dfrac{(\ )}{7}$
3. $+9$
4. $3(\ )$

Simplify each expression by canceling inverse operators. If no simpler form is possible, write "not possible."

5. $x - 3 + 3$

6. $\dfrac{5(a - b)}{5}$

7. $x + 6 - x$

8. $\dfrac{3n}{n}$

9. $\dfrac{x}{4} + 4$

10. $3 + y - 3$

11. $\dfrac{3a + b}{3}$

12. $\dfrac{x + 5}{y - 5}$

13. $\dfrac{(x + 3)6}{6}$

14. $\dfrac{24}{n} - n$

Maintain Your Skill

15. What number is five less than the product of three and eight? [3.1]

Name each form (sum, difference, product, or quotient) and then compute the basic numeral. [3.5]

16. $5 + \dfrac{3 \cdot 4}{2}$

17. $30 - 2(5 + 3)$

18. $\dfrac{8(7 - 3)}{2}$

19. $3\left(\dfrac{5 + 11}{4}\right)$

Write an equivalent operator in basic form, when this is defined. If this basic form is not yet defined, write "not defined." [4.2]

20. $3\left(\dfrac{(\ )}{15}\right)$

21. $\dfrac{3(\ )}{5}$

22. $\dfrac{14(\ )}{2}$

23. $\dfrac{\dfrac{(\ )}{6}}{3}$

24. $10\left(\dfrac{(\ )}{2}\right)$

25. $4(9(\ ))$

Chapter 4 Review

Construct a graphical model for the composition of the operators. [4.1, 4.2]

1. $\dfrac{\dfrac{(\ )}{3}}{2}$

2. $\dfrac{2(\ )}{6}$

Write a sentence that describes the action of the two operators. Then compute the basic form of the composite operator. [4.1, 4.2]

3. $\dfrac{\dfrac{(\ )}{6}}{2}$

4. $\dfrac{5(\ )}{35}$

5. $18\left(\dfrac{(\ )}{3}\right)$

6. $2(4(\ ))$

Interchange the order of the two operators. [4.3]

7. $x\left(\dfrac{(\ )}{y}\right)$

8. $\dfrac{\dfrac{(\ )}{x}}{y}$

An operand is given below, followed by operators to be used in the order stated. Construct this form and compute the basic numeral. [4.4]

9. $7;\ +8,\ 2(\ ),\ \dfrac{(\ )}{3}$

10. $5;\ +7,\ \dfrac{(\ )}{2},\ -3$

Analyze each expression by stating a basic operand and then listing a sequence of basic operators that can be joined one at a time to produce the given expression. If more than one operand and sequence of operators can be used, give the alternate possibilities. [4.5]

11. $\dfrac{21-5}{4}$

12. $x+\dfrac{n}{3}$

13. $a\left(\dfrac{12}{3}+8\right)$

14. $\dfrac{x-5}{n}+7$

Write the inverse of each operator. [4.6]

15. $+15$

16. $\dfrac{(\ )}{6}$

17. -11

18. $9(\ )$

Simplify each expression by canceling inverse operators. If no simpler form is possible, write "not possible." [4.6]

19. $12+x-12$

20. $\dfrac{3x}{x}$

21. $7-\dfrac{n}{7}$

22. $\dfrac{3a}{a-3}$

23. $\dfrac{(y-5)8}{8}$

24. $\dfrac{9-a+a}{a}$

25. $\dfrac{7x+6}{7x+3}$

26. $\dfrac{3w+4}{w+4}$

27. $\dfrac{n(2x+3)}{n}$

Chapter 5 — Solving Equations

5.1 – EQUIVALENT EQUATIONS: SUMS AND DIFFERENCES

OBJECTIVE

1. Identify and solve equations involving a sum and two addends.

An equation is a mathematical sentence stating that two amounts are equal. For a single equation, the equality sign "=" is used exactly once. Sometimes, however, we wish to combine several equations together. For example,

$$a = b = c \quad \text{or} \quad \begin{aligned} a &= b \\ &= c \end{aligned}$$

is a way of writing three equations: $a = b$, $b = c$, and $a = c$.

If there are no variables in an equation, then the equation will be either a true statement or a false one. For example,

$$3 + 4 = 7 \text{ is a true statement.}$$

$$3 + 4 = 5 \text{ is a false statement.}$$

On the other hand, equations containing variables may be true for some values of the variable and false for others. For this reason they are called **conditional equations**. For example, the equations

$$3 + 2 = x, \quad 3 + 2 = ?, \quad \text{and} \quad 3 + 2 = [\ ]$$

only represent a true statement on the condition that the placeholder be replaced by a numeral for 5.

In the equation $x = y$, the first or left member is x and the second or right member is y. If $x = y$ is to be true, it is required that x and y be different names for the same number.

The truth of an equation is not changed by switching the left and right members. This is known as the symmetric property of equality.

Symmetric Property of Equality

If $x = y$, then $y = x$.

The equations $x + y = z$ and $z - x = y$ are related to each other by the definition of subtraction (Lesson 1.5). That is, if three numbers are chosen for x, y and z so that one of the equations becomes a true statement, then the other equation will also be true for these values of x, y and z. For example, if $x = 2$, $y = 6$, and $z = 8$, then the statements

$$2 + 6 = 8 \quad \text{and} \quad 8 - 2 = 6$$

are both true. Likewise, if the three numbers should give a false statement for one equation, then the other equation will also be false. For example, if $x = 2$, $y = 5$, and $z = 8$, then the statements

$$2 + 5 = 8 \quad \text{and} \quad 8 - 2 = 5$$

are both false. Equations which have the same truth value as each other (either both true or both false) for any choice of the variables are said to be **equivalent equations**.

The equation $x + y = z$ is said to be solved **explicitly** for z. That is, z appears only once in the equation, and it is one of the members by itself. For example, the equation $3 + 4 = 7$ is explicit for 7, but not for 3 or 4.

To **solve** an equation for a variable (or a number) means to find an equivalent equation that is solved explicitly for the desired variable (or number).

We begin the process of solving equations by looking at the eight ways in which three numbers can be related by addition or subtraction. In Figure 5.1 we display these eight equations for the numbers 3, 5, and 8. The equations are arranged in a way to show how the equations are related to each other. The three properties that relate them are

s: symmetric property of equality

c: commutative property of addition

d: definition of subtraction

Figure 5.1

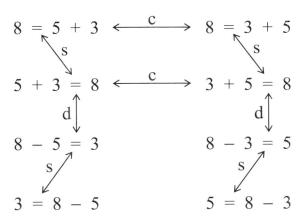

The top four equations in Figure 5.1 are solved explicitly for 8. The numbers 5 and 3 that are added together to get 8 are called **addends**. The top four equations all say that 8 is the sum of the two addends 5 and 3.

The lower four equations are each solved explicitly for one of the addends. The lower left pair are explicit for 3, and the lower right pair are explicit for 5. In each case we see that one addend is the sum minus the other addend.

We may summarize the two basic types of equations as follows:

Key Point

(S)   The sum equals one addend plus the other addend.

(See the top four equations in Figure 5.1.)

(A)   One addend equals the sum minus the other addend.

(See the bottom four equations in Figure 5.1.)

Example 1   Identify Sums and Addends

Identify the sum and the addends in each equation.

   (a) $4 + 7 = 11$ (b) $6 = 15 - 9$

(a)   The sum is 11 and the addends are 4 and 7.
(b)   The sum is 15 and the addends are 6 and 9.

Once we have identified the role of each number in an equation, any of the other equations can be written, as desired, simply by following the pattern of the appropriate basic type. We refer to this as the **3-number method** of solving an equation.

Example 2   Use the 3-number Method to Solve an Equation

Use the 3-number method to solve for each number in the equation that is not explicit, but do not compute.

   (a) $4 + 7 = 11$ (b) $6 = 15 - 9$

(a) $$4 + 7 = 11$$

We are asked to solve explicitly for 4 and for 7. We saw in Example 1 that they are both addends, and the sum is 11. To solve for the addend 4, we write the sum 11 minus the other addend 7.

$$4 = 11 - 7$$

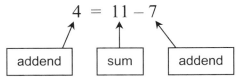

To solve for the addend 7, we write the sum 11 minus the other addend 4.

$$7 = 11 - 4$$

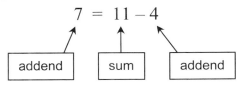

(b) $$6 = 15 - 9$$

We are asked to solve explicitly for 15 and 9. We saw in Example 1 that 15 is the sum and 9 is an addend. To solve for the sum 15, we add the two addends 9 and 6 together, in either order.

$$15 = 9 + 6 \quad \text{or} \quad 15 = 6 + 9$$

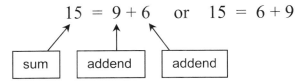

To solve for the addend 9, we write the sum 15 minus the other addend 6.

$$9 = 15 - 6.$$

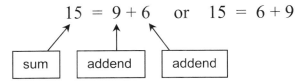

EXERCISE 5.1

DEVELOP YOUR SKILL

Identify the sum and the addends in each equation.

1. $12 = 9 + 3$ 2. $6 = 21 - 15$
3. $17 - 6 = 11$ 4. $13 + 5 = 18$
5. $4 + 12 = 16$ 6. $25 - 6 = 19$

Use the 3-number method to solve for each number in the equation that is not explicit, but do not compute. These are the same equations used in Exercises 1 – 6.

7. $12 = 9 + 3$ 8. $6 = 21 - 15$
9. $17 - 6 = 11$ 10. $13 + 5 = 18$
11. $4 + 12 = 16$ 11. $25 - 6 = 19$

MAINTAIN YOUR SKILL

13. This year Matt is twice as old as his younger brother. If Matt is 14 years old now, how old will his brother be next year? [3.2]

In Exercises 14 – 15, an operand number is given, followed by operators to be used in the order stated. Construct this form. If the form defines a counting number, then compute this number. If it does not, write "not defined." [4.4]

14. $15;\ +9,\ \dfrac{(\ )}{8},\ -1$ 15. $7;\ 3(\ ),\ -5,\ \dfrac{(\ )}{4}$

Analyze each expression by stating an operand and then listing a sequence of operators that can be joined one at a time to produce the given expression. If more than one operand and sequence of operators can be used, give the alternate possibilities. [4.5]

16. $7 + 3 \cdot 5$ 17. $5 + \dfrac{21}{3}$

18. $6(9 - 4) + 2$ 19. $7\left(\dfrac{6}{2} + 5\right)$

Simplify each expression by canceling inverse operators. If no simpler form is possible, write "not possible." [4.6]

20. $\dfrac{3x-2}{x-2}$

21. $\dfrac{4(n-3)}{n-3}$

22. $5+\dfrac{x}{5}-5$

23. $\dfrac{w+3}{w-3}$

24. $(y+7)\div 7$

25. $3\left(\dfrac{x}{3}\right)$

5.2 – SUMS AND DIFFERENCES WITH VARIABLES

OBJECTIVE

1. Solve sum and difference equations that contain variables.

    In Lesson 5.1 it was easy to identify the sum term in an equation: it was the largest of the three numbers. If an equation contains a variable, then it may not be obvious which term is the largest. But we can still identify the sum term by its role in the equation. Recall the two basic types of equations:

<u>Basic Types of Equations</u>

(S) The sum equals one addend plus the other addend.

[sum] = [one addend] + [other addend]

Example: $8 = 5 + 3$

(A) One addend equals the sum minus the other addend.

[one addend] = [sum] − [other addend]

Example: $5 = 8 - 3$

In Figure 5.2 we repeat the eight equations that relate sums and differences as in Figure 5.1, but this time using variables.

Figure 5.2

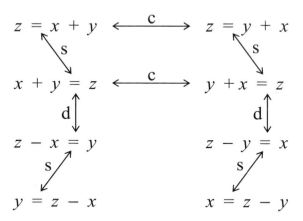

The top four equations state explicitly that

> z is the sum of the two addends x and y.

In the bottom four equations, it can be implied that z is the sum. For example, $z - x = y$ can be read as

> The sum z minus one addend x is the other addend y.

Study Tip:

This is why we call it the **3-number method** of solving an equation.

If an equation only involves adding and subtracting, we can think of it as a statement about *exactly three numbers*: a sum and two addends. Once these three terms have been identified, we can write any of the other seven equation forms by following the pattern of the appropriate basic type.

How do we figure out the role of each term? If one side of the equation shows addition, the sum term is on the other side:

> [one addend] + [other addend] = [**sum**]

> or   [**sum**] = [one addend] + [other addend]

If one side of the equation shows subtraction, the sum term is to the left of the minus sign.

$$[\textbf{sum}] - [\text{one addend}] = [\text{other addend}]$$

$$\text{or} \quad [\text{one addend}] = [\textbf{sum}] - [\text{other addend}]$$

Study Tip:

If "+" is on one side of the equation, the sum term is on the other side of the equation.

If "−" is on one side of the equation, the sum term is the term to the left of "−".

Example 1   Identify Sums and Addends

Identify the sum term and the two addends in each equation.

(a) $8 + m = t$ (b) $s = m - 4$

(c) $p - m = 7$ (d) $r = m + 2$

(a) We see the "+" on the left side. This means that t on the right side of the equation is the sum term. The addends are 8 and m. This is type (S).

(b) We see the "−" on the right side. This means that the m to the left of "−" is the sum term. The addends are s and 4. This is type (A).

(c) This is type (A) again. The sum term p minus the addend m is equal to the other addend 7.

(d) This follows type (S), where r is the sum, and the addends are m and 2.

Example 2   Solve an Equation

Solve each equation for m.

(a) $8 + m = t$ (b) $s = m - 4$

(c) $p - m = 7$ (d) $r = m + 2$

These are the same equations we had in Example 1. We have already identified m in each equation as the sum term or one of the addends. If m is the sum term, then it is equal to the sum of the addends. If m is one addend, then it is equal to the sum term minus the other addend.

(a) $$8 + m = t$$

m is an addend, so it is equal to the sum t minus the other addend 8: $m = t - 8$.

(b) $$s = m - 4$$

m is the sum, so it is the sum of the two addends s and 4: $m = s + 4$.

(c) $$p - m = 7$$

m is an addend. It equals the sum p minus the other addend 7: $m = p - 7$.

(d) $$r = m + 2$$

m is an addend and r is the sum. We have $m = r - 2$.

Study Tip:

This way of solving equations is very helpful in part (c). It would take several steps to solve it any other way.

If an equation involves several operations, sometimes the terms can be grouped together into one of the basic types of addition-subtraction equations.

Example 3　　Solve an Equation

Identify the sum term and then solve the equation for x.

(a) $3y = x - 7$　　　(b) $25 + x = \dfrac{m}{4}$

(a) We group $3y$ together as one number, and see that x is the sum of the two addends $3y$ and 7: $x = 3y + 7$.

(b) We think of $\dfrac{m}{4}$ as one number. It is the sum of the addends 25 and x. So the addend x is the sum $\dfrac{m}{4}$ minus the other addend 25: $x = \dfrac{m}{4} - 25$.

Example 4   Solve an Equation

Identify the sum term and then solve the equation for x.

   (a) $16 = 5y - x$ (b) $3n = x + \dfrac{m}{5}$ (c) $\dfrac{n+2}{y} = x - 3k$

(a) The "$-$" on the right side makes $5y$ the sum term. The addends are 16 and x. So the addend x is the sum $5y$ minus the other addend 16:

$$x = 5y - 16.$$

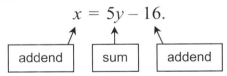

(b) The "$+$" on the right side makes $3n$ the sum term. The addend x equals the sum $3n$ minus the other addend $\dfrac{m}{5}$:

$$x = 3n - \dfrac{m}{5}$$

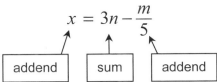

(c) We ignore the plus sign on the left side since it is inside the term $\dfrac{n+2}{y}$. The minus sign on the right side means that x is the sum term. We have $x = \dfrac{n+2}{y} + 3k$.

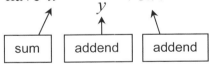

Remember:

[sum] = [one addend] + [other addend]

[one addend] = [sum] − [other addend]

EXERCISE 5.2

DEVELOP YOUR SKILL

Identify the sum term in each equation.

1. $a + x = 41$

2. $13 - x = y$

3. $32 = x - 5m$

4. $ab = 3 + x$

5. $\dfrac{a}{b} - x = 6$

6. $\dfrac{m}{n} = x - 8$

Solve each equation for x. These are the same equations used in Exercises 1 − 6.

7. $a + x = 41$

8. $13 - x = y$

9. $32 = x - 5m$

10. $ab = 3 + x$

11. $\dfrac{a}{b} - x = 6$

12. $\dfrac{m}{n} = x - 8$

Solve each equation for x and compute the basic numeral.

13. $x + 8 = 15$

14. $19 - x = 16$

15. $14 = x - 5$

16. $17 = x + 11$

17. $13 + x = 18$

18. $9 = 15 - x$

Maintain Your Skill

Do not compute, but write equivalent expressions with parentheses around a sum or a difference. Do not change the first term. When possible, give a second alternate form. [2.5]

19. $46 - 13 - 25$ **20.** $32 - 7 + 20$

21. $17 + 21 - 14$ **22.** $25 + 17 - 23$

23. $24 - 18 + 6$ **24.** $15 + 20 + 6$

5.3 – Equivalent Equations: Products and Quotients

OBJECTIVE

1. Identify and solve equations involving a product and two factors.

The study of equivalent equations for products and quotients is similar to that for sums and differences. We now use the commutative property of multiplication (instead of addition), and use the definition of division instead of that for subtraction.

s: symmetric property of equality

c: commutative property of multiplication

d: definition of division

Figure 5.3

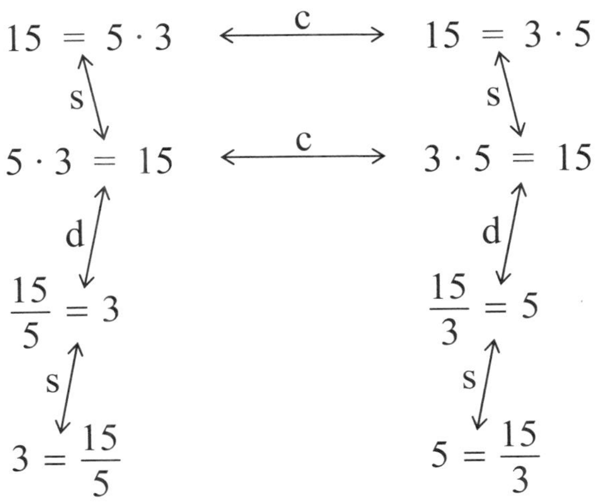

$$15 = 5 \cdot 3 \quad \xleftarrow{\ \ c\ \ } \quad 15 = 3 \cdot 5$$

$$5 \cdot 3 = 15 \quad \xleftarrow{\ \ c\ \ } \quad 3 \cdot 5 = 15$$

$$\frac{15}{5} = 3 \qquad\qquad \frac{15}{3} = 5$$

$$3 = \frac{15}{5} \qquad\qquad 5 = \frac{15}{3}$$

In these equations, the 5 and the 3 are **factors** and 15 is the **product**. When numbers greater than 1 are used, then the product is always larger than either of the two factors. In the upper four equations, the product is shown explicitly as one factor times the other factor. In the lower four equations, the product is not explicit. In these equations, one of the factors is equal to the product divided by the other factor.

We may summarize the two basic types of equations as follows:

Key Point

(P) The product equals one factor times the other factor.

$$[\textbf{product}] = [\text{one } \textbf{factor}] \times [\text{other } \textbf{factor}]$$

(See the top four equations in Figure 5.3.)

(F) One factor equals the product divided by the other factor.

$$[\text{one } \textbf{factor}] = \frac{[\textbf{product}]}{[\text{other } \textbf{factor}]}$$

(See the bottom four equations in Figure 5.3.)

The 3-number method may be used to transform equations built with products and quotients in a manner similar to that used in Lessons 5.1 and 5.2 with sums and differences. In this case we begin by identifying which term is the product and which terms are the two factors.

Example 1   Solve an Equation

Solve for each number that is not explicit, but do not compute.

(a) $4 \cdot 9 = 36$ (b) $5 = 35 \div 7$

(a) The equation is solved explicitly for the product 36. To solve for the factor 4, we write the product 36 divided by the other factor 9.

$$\boxed{\text{factor}} \longrightarrow 4 = \frac{36 \quad \longleftarrow \boxed{\text{product}}}{9 \quad \longleftarrow \boxed{\text{factor}}}$$

To solve for the factor 9, we write the product 36 divided by the other factor 4.

$$\boxed{\text{factor}} \longrightarrow 9 = \frac{36 \quad \longleftarrow \boxed{\text{product}}}{4 \quad \longleftarrow \boxed{\text{factor}}}$$

(b) The equation $5 = 35 \div 7$ is solved explicitly for the factor 5. To solve for the product 35, we multiply the two factors 5 and 7, in either order.

$$35 = 5 \cdot 7 \quad \text{or} \quad 35 = 7 \cdot 5$$

$$\boxed{\text{product}} \quad \boxed{\text{factor}} \quad \boxed{\text{factor}}$$

To solve for the factor 7, we write the product 35 divided by the other factor 5.

$$\boxed{\text{factor}} \longrightarrow 7 = \frac{35 \quad \longleftarrow \boxed{\text{product}}}{5 \quad \longleftarrow \boxed{\text{factor}}}$$

We could also write our answer horizontally as
$$7 = 35 \div 5.$$

EXERCISE 5.3

DEVELOP YOUR SKILL

Identify the product term and the two factors in each equation.

1. $24 = 3 \cdot 8$

2. $4 = \dfrac{20}{5}$

3. $\dfrac{18}{2} = 9$

4. $9 \cdot 5 = 45$

5. $8 \cdot 12 = 96$

6. $\dfrac{56}{7} = 8$

Use the 3-number method to solve for each number that is not explicit, but do not compute. These are the same equations used in Exercises 1 – 6.

7. $24 = 3 \cdot 8$

8. $4 = \dfrac{20}{5}$

9. $\dfrac{18}{2} = 9$

10. $9 \cdot 5 = 45$

11. $8 \cdot 12 = 96$

12. $\dfrac{56}{7} = 8$

MAINTAIN YOUR SKILL

Solve each equation by using an invariant principle. [1.9, 3.9]

13. $348 + 257 = 346 + x$

14. $531 - x = 521 - 237$

15. $\dfrac{18}{6} = \dfrac{x}{12}$

16. $x \cdot 4 = 25 \cdot 20$

Identify the sum term in each equation. [5.2]

17. $2n + x = 53$

18. $a = x - 10$

19. $37 - x = 3k$

20. $y = 8 + x$

Solve each equation for x. These are the same equations used in Exercises $17 - 20$.
[5.2]

21. $2n + x = 53$ **22.** $a = x - 10$

23. $37 - x = 3k$ **24.** $y = 8 + x$

5.4 – PRODUCTS AND QUOTIENTS WITH VARIABLES

OBJECTIVE

1. Solve product and quotient equations that contain variables.

    In Lesson 5.3 it was easy to identify the product term in an equation: it was the largest of the three numbers. If an equation contains a variable, then it may not be obvious which term is the largest. But we can still identify the product term by its role in the equation. Recall the two basic types of equations:

<u>Basic Types of Equations</u>

(P) The product equals one factor times the other factor.

$$[\text{product}] = [\text{one factor}] \times [\text{other factor}]$$

$$\text{Example: } 15 = 5 \cdot 3$$

(F) One factor equals the product divided by the other factor.

$$[\text{one factor}] = \frac{[\text{product}]}{[\text{other factor}]}$$

Or, written horizontally,

$$[\text{one factor}] = [\text{product}] \div [\text{other factor}]$$

$$\text{Example: } 5 = \frac{15}{3} \text{ or } 5 = 15 \div 3$$

    In Figure 5.4 we repeat the eight equations that relate products and quotients as in Figure 5.3, but this time using variables.

Figure 5.4

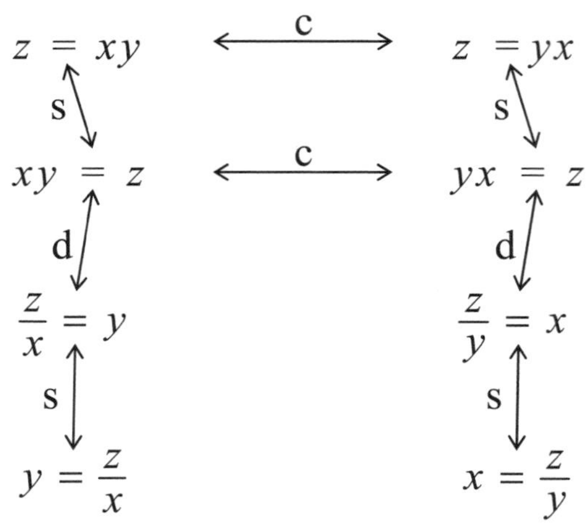

$$z = xy \quad \xleftarrow{\quad c \quad}\rightarrow \quad z = yx$$

$$\downarrow s \qquad\qquad\qquad \downarrow s$$

$$xy = z \quad \xleftarrow{\quad c \quad}\rightarrow \quad yx = z$$

$$\nearrow d \qquad\qquad\qquad \nearrow d$$

$$\frac{z}{x} = y \qquad\qquad \frac{z}{y} = x$$

$$\downarrow s \qquad\qquad\qquad \downarrow s$$

$$y = \frac{z}{x} \qquad\qquad x = \frac{z}{y}$$

The top four equations state explicitly that

z is the product of the two factors x and y.

In the bottom four equations, it can be implied that z is the

product. For example, $\dfrac{z}{x} = y$ can be read as

The product z divided by one factor x is the other factor y.

When we look at a multiplying/dividing equation, how do we figure out the role of each part of the equation? If two terms are being multiplied on one side of the equation, the product term is on the other side:

[one factor] × [other factor] = [**product**]

or   [**product**] = [one factor] × [other factor]

If two terms are being divided on one side of the equation, the product term is the dividend (top) of the quotient,

$$\frac{[\textbf{product}]}{[\text{one factor}]} = [\text{other factor}]$$

or $[\text{one factor}] = \dfrac{[\textbf{product}]}{[\text{other factor}]}$

Or, if the division equation is written horizontally, the product term is to the left of the "÷" symbol.

$$[\text{one factor}] = [\mathbf{product}] \div [\text{other factor}]$$

or $\quad [\mathbf{product}] \div [\text{other factor}] = [\text{one factor}]$

Study Tip:

If there is multiplying on one side of the equation, the product term is the other side of the equation.

If there is dividing on one side of the equation, the product term is the dividend in the quotient.

Example 1   Identify Products and Factors

Identify the product term and the two factors in each equation.

(a) $\;8m = t$

(b) $\;s = \dfrac{m}{4}$

(c) $\;p \div m = 7$

(d) $\;a + b = (m)(2)$

(a)   The multiplying on the left side makes t the product. The factors are 8 and m. This is type (P).

(b)   The dividing on the right side makes m the product. The factors are s and 4. This is type (F).

(c)   The dividing on the left makes p the product (of the factors m and 7).

(d)   View $a + b$ as one number. It is the product (of the factors m and 2.)

Example 2   Solve an Equation

Solve each equation for m.

(a) $\;8m = t$

(b) $\;s = m \div 4$

(c) $\;\dfrac{p}{m} = 7$

(d) $\;a + b = (m)(2)$

These are the same equations we had in Example 1. We have already identified m in each equation as the product term or one of the factors. If m is the product term, then it is equal to the product of the two factors. If m is one factor, then it is equal to the product divided by the other factor.

(a) $$8m = t$$

Since m is a factor, and it equals the product t divided by the other factor 8:

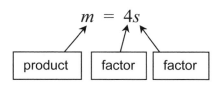

$$\boxed{\text{factor}} \longrightarrow m = \frac{t}{8}$$

with labels: product (pointing to t), factor (pointing to 8)

(b) $$s = m \div 4$$

This time, m is the product and the factors are 4 and s. To solve for m, we multiply the two factors together:

$$m = 4s$$

product   factor   factor

(c) $$\frac{p}{m} = 7$$

Since m is a factor, it is equal to the product p divided by the other factor 7:

$$\boxed{\text{factor}} \longrightarrow m = \frac{p}{7}$$

with labels: product (pointing to p), factor (pointing to 7)

(d) $\qquad\qquad a + b = (m)(2)$

Once again, m is a factor. So m equals the product $(a + b)$ divided by the other factor 2. We may write this as

$$m = \frac{a+b}{2} \qquad \text{or} \qquad m = (a + b) \div 2.$$

This is really important!

Remember:

$$[\text{product}] = [\text{one factor}] \times [\text{other factor}]$$

$$[\text{one factor}] = \frac{[\text{product}]}{[\text{other factor}]}$$

EXERCISE 5.4

DEVELOP YOUR SKILL

Identify the product term and the two factors in each equation.

1. $5y = n$

2. $24 = \dfrac{x}{y}$

3. $8 = \dfrac{y}{a+3}$

4. $y(b - 5) = 48$

5. $\dfrac{n+6}{y} = m$

6. $c - 3 = \dfrac{y}{n}$

Solve each equation for y. These are the same equations used in Exercises 1 – 6.

7. $5y = n$

8. $24 = \dfrac{x}{y}$

9. $8 = \dfrac{y}{a+3}$

10. $y(b-5) = 48$

11. $\dfrac{n+6}{y} = m$

12. $c-3 = \dfrac{y}{n}$

Solve each equation for x and compute the basic numeral.

13. $x \cdot 4 = 12$

14. $\dfrac{24}{x} = 8$

15. $14 = \dfrac{x}{2}$

16. $10 = x \cdot 2$

17. $6 \cdot x = 24$

18. $8 = \dfrac{40}{x}$

MAINTAIN YOUR SKILL

19. Sally planted 4 rows of green beans in her garden with 24 plants in each row. If she arranged the same number of plants in 8 rows, let x be the number of plants in each row. Since the total number of plants does not change, the number of rows times the number of plants remains the same. This gives us the equation $4 \cdot 24 = 8 \cdot x$. Use an invariant principle to solve this equation and find how many plants would be in each row if there were 8 rows. [5.3]   12 plants

Interchange the order of the two operators. [2.4, 4.3]

20. $\dfrac{4(\ )}{20}$

21. $\dfrac{\frac{(\ )}{8}}{2}$

22. $-r - s$

23. $-u + v$

24. $a\left(\dfrac{(\ )}{b}\right)$

25. $\dfrac{x(\ )}{y}$

5.5 – REMOVING INCREASES AND DECREASES

OBJECTIVE

1. Solve an equation by removing increases and decreases.

     The problem of simplifying an equation or solving explicitly for a particular variable is central to the study of algebra. In Lessons 5.1 – 5.4 we introduced a method based on identifying a sum or a product relationship within the equation. In this lesson and the next we use the concept of operators to develop an additional way of solving equations.

     To solve an equation such as $x - 5 = 7$, we seek to simplify the equation so that the left side is just x by itself. The expression $x - 5$ can be built by starting with the operand x and joining the operator -5, a decrease of 5..

$$x \xrightarrow{\text{join } -5} x - 5$$

Solving the equation can be thought of as reversing this process:

$$x - 5 \xrightarrow{\text{remove } -5} x$$

     The effect of removing (and joining) operators can be seen by comparing the values before and after the changes. Let's look at a couple of examples.

Example 1   **Effect of Joining or Removing Operators**

Suppose John has 4 dimes and Frank gives him 5 more dimes. We see that 4 increases to 9 by joining the operator $+5$. On the other hand, if Frank decides not to give him 5 dimes, then the 9 that John might have had remains at 4. We see that 9 (written as $4 + 5$) decreases to 4 by removing the operator $+5$.

Change	Description
$4 \longrightarrow 4 + 5$	4 is increased by 5 by joining the operator $+5$.
$4 + 5 \longrightarrow 4$	9 is decreased by 5 by removing the operator $+5$.

> ### Example 2   Effect of Joining or Removing Operators
>
> Suppose Judy has 8 dimes and gives 5 dimes to Ruth. We see that 8 decreases to 3 by joining the operator -5. On the other hand, if Judy decides not to give away 5 dimes, then the 3 that she would have had left increases to the 8 she started with. We see that 3 (written as $8 - 5$) increases to 8 by removing the operator -5.
>
> | Change | Description |
> |---|---|
> | $8 \longrightarrow 8 - 5$ | 8 is decreased by 5 by joining the operator -5. |
> | $8 - 5 \longrightarrow 8$ | 3 is increased by 5 by removing the operator -5. |

By comparing Examples 1 and 2 we see that there are two ways to increase by 5: we can join $+5$ or remove -5. Similarly, there are two ways to decrease by 5: we can join -5 or remove $+5$.

In general, removing an operator makes the same change as joining the inverse operator of the same magnitude.

Joining $+x$ or $x+$ results in an increase of x.

Removing $-x$ results in an increase of x.

Joining $-x$ results in a decrease of x.

Removing $+x$ or $x+$ results in a decrease of x.

Study Tip:

This is called the **operator method** of solving an equation.

If we have an equation and we make equivalent changes to both sides of the equation, then the truth value of the equation will not be changed. That is, the original equation and the equation after the change will be equivalent. We can make these changes by joining operators or removing operators. In fact, we can remove an operator from one side and join the inverse operator to the other side, since both changes have exactly the same effect. We call this the **operator method** of solving an equation.

Key Point

Two equations are equivalent if one can be changed into the other by removing an operator from one side and then joining the inverse operator to the other side.

Study Tip:

When you remove an operator from one side of an equation, you join the *inverse* operator to the other side of the equation.

Example 3   **Remove an Operator and Join its Inverse**

Compare $x + 3 = 5$

with $x = 5 - 3.$

Both sides of the first equation are decreased by 3. The left side is decreased by removing $+ 3$ and the right side by joining $- 3$. The two equations are equivalent.

Example 4   **Solve an Equation**

Solve each equation for x by using the operator method. Then check your answer.

    (a) $x + 5 = 8$ (b) $9 = x - 4$

(a)   Remove the increase of 5 from the left side and join its inverse, a decrease of 5, to the right side:

Study Tip:

You can think of subtracting 5 from both sides of the equation.

You subtract 5 from the left side by removing $+ 5$.

You subtract 5 from the right side by joining $- 5$.

$x + 5 = 8$ Write the equation.

$x = 8 - 5$ Remove + 5 from the left side
                     and join − 5 to the right side.

$x = 3$ Compute 8 − 5.

To check our answer, we substitute $x = 3$ in the original equation to see if we get a true sentence. When $x = 3$, the original equation $x + 5 = 8$ becomes $3 + 5 = 8$, which is true.

(b) Remove the decrease of 4 from the right side and join its inverse, an increase of 4, to the left side.

$$9 = x - 4 \qquad \text{Write the equation.}$$

$$9 + 4 = x \qquad \text{Remove } -4 \text{ from the right side and join } +4 \text{ to the left side.}$$

$$13 = x \qquad \text{Compute } 9 + 4.$$

It is customary to write the letter or number we are solving for on the left side of the equation in our answer, so we use the symmetric property of equality to do this:

$$x = 13 \qquad \text{Use the symmetric property of equality.}$$

When we substitute 13 for x in the original equation we get $9 = 13 - 4$, which is true. So our answer checks.

EXERCISE 5.5

DEVELOP YOUR SKILL

Solve each equation for x by using the operator method. Compute the basic numeral and then check your answer.

1. $x + 3 = 15$ 2. $x - 5 = 8$

3. $18 = x - 10$ 4. $7 + x = 13$

5. $9 = 4 + x$ 6. $x - 6 = 5$

Solve each equation for x by using the operator method.

7. $n = x + 4$ 8. $10 = x - n$

9. $x + 2y = 6$ 10. $\dfrac{a}{b} = x - 3$

11. $x - 3y = cd$ 12. $y = x + 4m$

MAINTAIN YOUR SKILL

13. Each cookie contains 6 chocolate chips. How many chocolate chips would be in 75 cookies? [3.1]

Write the corresponding equation that is matched to it by the definition of subtraction or the definition of division. [1.5, 3.2]

14. $x - y = m$

15. $a + k = s$

16. $cx = t$

17. $\dfrac{x}{a} = n$

Solve each equation by using an invariant principle. [3.8, 3.9]

18. $x \cdot 8 = 42 \cdot 16$

19. $\dfrac{24}{x} = \dfrac{48}{16}$

Simplify each expression by canceling inverse operators. If no simpler form is possible, write "not possible." [4.6]

20. $\dfrac{a(b+c)}{a}$

21. $\dfrac{xy}{x} + \dfrac{yz}{y}$

22. $\dfrac{ab - c}{a}$

23. $\dfrac{a+b}{a-b}$

24. $\dfrac{ab}{b} - b$

25. $\dfrac{b(c-d)}{b} + d$

5.6 – REMOVING EXPANSIONS AND CONTRACTIONS

OBJECTIVE

1. Solve an equation by removing expansions and contractions.

   Since expansions and contractions are inverse operations, we would expect that removing an expansion would have the same effect as joining a contraction. Let's look at a couple of examples.

Example 1   Effect of Joining or Removing Operators

Beth is collecting bags of pinecones to make a wreath. Each bag contains 5 pinecones. If she collects 4 bags, then she will have a total of 20 pinecones. We see that 5 expands to 20 by joining 4( ). On the other hand, if she gets tired after collecting only one bag, then the 20 she might have had only amounts to 5. We see that 20 (written as $4 \cdot 5$) is contracted to 5 by removing the 4( ).

Change	Description
$5 \longrightarrow 4(5)$	5 is expanded by 4 by joining the operator 4( ).
$4(5) \longrightarrow 5$	20 is contracted by 4 by removing the operator 4( ).

Example 2   Effect of Joining or Removing Operators

Scott wants to make a ladder to reach his tree house. He has a board 8 feet long that he decided to cut into 4 equal pieces to nail on the tree. Each piece will be 2 feet long. We see that 8 is contracted to 2 by joining $\frac{(\ )}{4}$. On the other hand, if he decided not to cut his board, then it will remain 8 feet long. So 2 (written as $\frac{8}{4}$) is expanded to 8 by removing the operator $\frac{(\ )}{4}$.

Change	Description
$8 \longrightarrow \frac{8}{4}$	8 is contracted by 4 by joining the operator $\frac{(\ )}{4}$.
$\frac{8}{4} \longrightarrow 8$	2 is expanded by 4 by removing the operator $\frac{(\ )}{4}$.

In general, if $x \neq 0$, multiplication by x results if the operators $x(\ )$ or $(\ )x$ are joined or if $\dfrac{(\ )}{x}$ is removed. Division by x results if the operator $\dfrac{(\ )}{x}$ is joined or if the operators $x(\ )$ or $(\ )x$ are removed. Of course, we do not want to join or remove the operator $0(\ )$, since it has no inverse.

In solving an equation, we want to make equivalent changes to both sides of the equation. So if we remove an operator from one side of an equation, we want to join its inverse to the other side.

For example, compare the two equations:

$$\frac{x}{2} = 4 \quad \text{and} \quad x = 4 \cdot 2$$

Both sides of the first equation have been multiplied by 2. The left side is multiplied by 2 by removing the contraction operator $\dfrac{(\ )}{2}$, and the right side is multiplied by 2 by joining the expansion operator $(\ )2$. The two equations are equivalent.

In the same way, compare these two equations:

$$6x = 24 \quad \text{and} \quad x = \frac{24}{6}.$$

This time, both members of the first equation have been divided by 6. The left member is divided by 6 by removing the expansion operator $6(\ )$, and the right member is divided by 6 by joining the contraction operator $\dfrac{(\ )}{6}$. The two equations are equivalent.

Example 3   **Solve an Equation by Removing Operators**

Solve each equation for x.

(a) $4x = 20$ (b) $c = \dfrac{x}{2}$

(a) We wish to have x by itself, so we want to remove the expansion by 4. This has the effect of dividing the left side by 4. On the right side we join a contraction by 4:

$$4x = 20 \quad \Rightarrow \quad x = \frac{20}{4}.$$

Now we compute and get $x = 5$.

(b) This time we want to remove contraction by 2 from the right side. This has the effect of multiplying by 2, so we join an expansion by 2 to the left side.

$$c = \frac{x}{2} \quad \Rightarrow \quad 2c = x.$$

Finally, we use the symmetric property of equality to interchange the left and right sides of the equation. This puts the x on the left side:

$$x = 2c$$

When operators are used with equations, we think of each side of the equation as representing just *one* number. So when operators are joined to one side of an equation, the change must be to that side as a whole. For expansions, the parentheses for $n(\ )$ or $(\ )n$ must enclose *all* of the parts to which they are applied. Likewise for contractions, the bar of the division operator $\dfrac{(\ )}{n}$ is written underneath the *entire* expression.

Example 4   **Solve an Equation by Removing Operators**

Solve each equation for x.

(a) $\dfrac{x}{3} = a + k$ (b) $b + 4y = nx$ (c) $x + 3 = \dfrac{h}{w}$

(a) We want to remove the contraction $\dfrac{(\;)}{3}$ from the left side and join its inverse $3(\;)$ to the right side. We put the parentheses around the *whole* expression on the right:

$$x = 3(a + k)$$

(b) We want to remove the expansion $n(\;)$ from the right side and join its inverse $\dfrac{(\;)}{n}$ to the left side. The division bar goes under the *whole* expression on the left side:

$\dfrac{b + 4y}{n} = x$. Finally, rewrite the equation with the x on the left side:

$$x = \dfrac{b + 4y}{n}$$

(c) To get x by itself, we want to remove the increase of 3 from the left side and join its inverse, a decrease of 3, to the right side:

$$x = \dfrac{h}{w} - 3$$

Study Tip:

When we join -3 to the right side, we join it to the *whole* expression on the right. We do **not** get

$x = \dfrac{h - 3}{w}$

or $x = \dfrac{h}{w - 3}$.

EXERCISE 5.6

DEVELOP YOUR SKILL

Solve each equation for x by removing and joining operators. Compute the basic numeral when possible.

1. $3x = 12$

2. $8 = \dfrac{x}{4}$

3. $m = 5x$

4. $8x = 40$

5. $y = \dfrac{x}{5}$

6. $\dfrac{x}{a} = n - 1$

7. $t = x(m + 3)$

8. $bx = c + 7$

9. $x - 5 = \dfrac{n}{3}$

10. $3y = 7 + x$

11. $n + 4 = 7x$

12. $\dfrac{x}{2} = b - a$

MAINTAIN YOUR SKILL

Construct graphical models for the composition of the operators. [2.4, 4.2]

13. $-2 + 8$

14. $-3 + 6$

15. $8\left(\dfrac{(\ )}{2}\right)$

16. $6\left(\dfrac{(\ )}{3}\right)$

Name each form (sum, difference, product, or quotient) and then compute the basic numeral. [3.5]

17. $14 - 3 \cdot 2$

18. $3(8 - 2)$

19. $5 + 2 \cdot 6$

20. $4 \cdot 6 - 2$

21. $\dfrac{3 \cdot 8}{4}$

22. $3\left(\dfrac{8}{4}\right)$

23. $4 \cdot 5 + \dfrac{6}{2}$

24. $\dfrac{4 \cdot 5 + 6}{2}$

5.7 – MIXED EQUATIONS

OBJECTIVE

1. Solve an equation by the 3-number method or by removing operators.

In this chapter we have studied two different ways of solving equations. The first method sees three numbers in the equation: either a sum and two addends, or a product and two factors. The given equation is always explicit for one of these three numbers. Two more explicit equations can then be written using these same three numbers. This is called the **3-Number Method**.

The second method makes changes by removing an operator from one side of the equation and joining the inverse operator to the other side. This has the effect of making an equivalent change to both sides. This is called the **Operator Method**.

In any particular problem, we have a choice of which method to use. Often it is helpful to work a problem both ways and use the alternate method as a check on our work.

Example 1   **Solve an Equation**

Solve the equation for x.

$$x + 7 = a$$

We can see x as an addend. To solve for it we write the sum a minus the other addend 7:

$$x = a - 7$$

Or, we can subtract 7 from both sides of the original equation. We remove the increase of 7 from the left side and join a decrease of 7 to the right side. We get the same result.

Example 2   Solve an Equation

Solve the equation for x.

$$n - x = 5$$

If we were to remove an operator from the left side, we would have to remove the decrease of x.   But this would not help us solve for x. So the best way to solve this equation for x is with the 3-number method.

We see x as an addend. To solve for it, we write the sum n minus the other addend 5:

$$x = n - 5$$

Some equations can be viewed in more than one way using the 3-number method. For example, consider the equation

$$\frac{a}{b} = c + d.$$

If we group the right side together as one number and see

$$\frac{a}{b} = (c + d),$$

then a is the product term, and b and $c + d$ are the factors. This would be a useful way to look at the equation if we were asked to solve for a or b.

But if we group the left side together as one number and see

$$\left(\frac{a}{b}\right) = c + d,$$

then $\frac{a}{b}$ is the sum of the two addends c and d. Viewing the equation this way makes it easy to solve for c or d.

Example 3   **Identify the Sum or the Product**

View each equation as consisting of exactly three parts. Identify the part that is the sum (S) or the product (P) of the other two parts. Give all possible interpretations.

(a) $m - 3 = \dfrac{n}{4}$ (b) $2w = a + 5$

(a)   Grouping the left side together as one number makes n the product.

$$(m - 3) = \frac{n}{4} \longleftarrow \boxed{\text{product term}}$$

Grouping the right side together as one number makes m the sum.

$$\boxed{\text{sum term}} \searrow$$
$$m - 3 = \left(\frac{n}{4} \right)$$

We may write our answer as (P): n and (S): m.

(b) Grouping the left side together makes it the sum:

$$\boxed{\text{sum term}} \searrow$$
$$(2w) = a + 5$$

Grouping the right side together and seeing the multiplying on the left side makes $a + 5$ the product:

$$\boxed{\text{product term}}$$
$$(2)(w) = (a + 5)$$

We may write the answer as (S): $2w$ and (P): $a + 5$.

EXERCISE 5.7

DEVELOP YOUR SKILL

Solve each equation for x either by the Operator Method or by the 3-Number Method. If you solve using the Operator Method, indicate which operator was removed. If you solve by the 3-Number Method, indicate the role that x plays in the equation (sum, addend, product, or factor).

1. $x - 3 = k$

2. $2x = y$

3. $\dfrac{n}{x} = 7$

4. $20 - x = m$

5. $a + x = 18$

6. $a + b = \dfrac{x}{3}$

View each equation as consisting of exactly three parts. Identify the part that is the sum (S) or the product (P) of the other two parts. Give all possible interpretations.

7. $x + 3 = \dfrac{a}{5}$

8. $3b = x - 6$

9. $n - 7 = \dfrac{x}{4}$

10. $\dfrac{a}{x} = y + 2$

11. $a + 5 = y - x$

12. $cy = \dfrac{n}{x}$

Solve each equation for x. These are the same equations used in Exercises 7 – 12.

13. $x + 3 = \dfrac{a}{5}$

14. $3b = x - 6$

15. $n - 7 = \dfrac{x}{4}$

16. $\dfrac{a}{x} = y + 2$

17. $a + 5 = y - x$

18. $cy = \dfrac{n}{x}$

MAINTAIN YOUR SKILL

Becky has 23 stamps and Francis has 18 stamps. Use this information in Exercises 19 and 20.

19. How many stamps do Becky and Francis have together? [1.1]

20. How many more stamps does Becky have than Francis? [1.5]

Use distributive principles to replace each product or quotient by an equivalent sum or difference, and to replace each sum or difference by an equivalent product or quotient. Do <u>not</u> compute and do not use the commutative principle. [3.7]

21. $3n + 3k$

22. $c(x - 4)$

23. $7y + 3y$

24. $\dfrac{m + 6}{r}$

25. $\dfrac{a}{5} - \dfrac{b}{5}$

26. $(x + 2)m$

CHAPTER 5 REVIEW

Solve each equation for the number 26, but do not compute. [5.1, 5.3]

1. $71 = 45 + 26$

2. $26 - 17 = 9$

3. $54 - 26 = 28$

4. $78 = 3 \cdot 26$

5. $\dfrac{26}{2} = 13$

6. $\dfrac{156}{26} = 6$

View each equation as a sum and two addends. Identify the part that is playing the role of the sum. [5.2]

7. $xy = ab + \dfrac{c}{d}$

8. $am - y = \dfrac{x}{n}$

View each equation as a product and two factors. Identify the part that is playing the role of the product. [5.4]

9. $xy = ab + \dfrac{c}{d}$

10. $am - y = \dfrac{x}{n}$

Solve each equation for x by removing and joining operators. [5.5, 5.6]

11. $x - 5 = n$

12. $3x = h$

13. $c = \dfrac{x}{7}$

14. $y = x + 6$

Solve each equation for x. [5.7]

15. $\dfrac{m}{2} = x - 3$

16. $2x = 3n + 1$

17. $5y - 2 = \dfrac{n}{x}$

18. $6k - x = a$

19. $\dfrac{x}{c} = m + 3$

20. $2y = x + 7$

21. $y - 5 = ax$

22. $x - n = 6h$

23. $\dfrac{n + 3}{x} = y$

24. $15 = 2m - x$

Chapter 6 – Powers

6.1 – EXPONENTS

In Chapter 3 we showed how multiplication can be seen as repeated addition. In the product $3 \cdot 5$, we start with 0 and add 5 a total of 3 times:

$$3 \cdot 5 = 0 + 5 + 5 + 5$$

To show repeated multiplication, we now introduce a useful form called **exponentiation**, or more commonly, **raising to a power**.

For example, 2 to the third power is written this way:

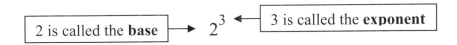

2 is called the **base** $\rightarrow$ 2^3 $\leftarrow$ 3 is called the **exponent**

To compute the basic numeral, we start with 1 and multiply by 2 a total of 3 times:

$$2^3 = 1 \cdot 2 \cdot 2 \cdot 2$$

The fourth power of 5 is written like this:

5 is the **base** $\rightarrow$ 5^4 $\leftarrow$ 4 is the **exponent**

To compute the basic numeral, we start with 1 and multiply by 5 a total of 4 times:

$$5^4 = 1 \cdot 5 \cdot 5 \cdot 5 \cdot 5$$

In general, the expression b^n is called a **power**, b is called the **base**, and n is called the **exponent**.

To compute the power b^n, we begin with 1 and multiply by b a total of n times. This means that the exponent counts the number of times that the unit 1 is multiplied by the base. Or, in the language of operators, the exponent n counts the number of times that the expansion ( )b is joined to 1:

$$b^n = 1 \cdot \underbrace{b \cdot b \cdot b \cdot \ldots \cdot b}_{n \text{ times}}$$

Example 1   **An Operator Model for Powers**

Power Form	Number of Expansions	Operator Model	Description	Basic Numeral
3^0	none	1	1 is multiplied by 3, 0 times	1
3^1	one	$1 \cdot 3$	1 is multiplied by 3, 1 time	3
3^2	two	$1 \cdot 3 \cdot 3$	1 is multiplied by 3, 2 times	9
3^3	three	$1 \cdot 3 \cdot 3 \cdot 3$	1 is multiplied by 3, 3 times	27
3^4	four	$1 \cdot 3 \cdot 3 \cdot 3 \cdot 3$	1 is multiplied by 3, 4 times	81

Compare this with Example 1 in Lesson 3.1.

Let's compute several more powers and see what patterns we can find.

$$3^3 = 1 \cdot 3 \cdot 3 \cdot 3 = 27 \qquad 3^2 = 1 \cdot 3 \cdot 3 = 9 \qquad 3^1 = 1 \cdot 3 = 3 \qquad 3^0 = \mathbf{1}$$

$$2^3 = 1 \cdot 2 \cdot 2 \cdot 2 = 8 \qquad 2^2 = 1 \cdot 2 \cdot 2 = 4 \qquad 2^1 = 1 \cdot 2 = 2 \qquad 2^0 = \mathbf{1}$$

$$1^3 = 1 \cdot 1 \cdot 1 \cdot 1 = 1 \qquad 1^2 = 1 \cdot 1 \cdot 1 = 1 \qquad 1^1 = 1 \cdot 1 = 1 \qquad 1^0 = \mathbf{1}$$

$$0^3 = 1 \cdot 0 \cdot 0 \cdot 0 = \mathbf{0} \qquad 0^2 = 1 \cdot 0 \cdot 0 = \mathbf{0} \qquad 0^1 = 1 \cdot 0 = \mathbf{0} \qquad \boxed{}$$

Study Tip:

0^0 is not defined.

We notice that the value of each power in the bottom row is zero. And the value of each power in the right column is one. But the entry for 0^0 is omitted. If we were to decide that $0^0 = 1$, this would be an exception in the bottom row, since all other powers are 0 when the base is 0. But if we decide that $0^0 = 0$, this would be an exception in the right column, since all other powers are 1 when 0 is the exponent. This conflict can be avoided by not giving a meaning to 0^0.

We have the following general rules:

Key Point

$$\text{For all } b, \ b^1 = b. \qquad \text{For all } b \neq 0, \ b^0 = 1.$$

$$\text{For all } n \neq 0, \ 0^n = 0.$$

Example 2   Compute a Power

Compute the following powers.

(a) 6^2 (b) 2^4 (c) 5^0

(a) By definition, $6^2 = 1 \cdot 6 \cdot 6 = 36$. When the exponent is greater than 1, computing a power can be shortened by dropping the $1 \cdot$ at the front. So we may compute 6^2 as $6 \cdot 6 = 36$. It is important, however, to keep the initial term 1 when computing a power having zero as an exponent.

(b) We may compute 2^4 as $2 \cdot 2 \cdot 2 \cdot 2 = 16$.

(c) Since the exponent is 0, we have to use the definition. So we start with 1 and multiply by 5 no times to just have 1. That is, $5^0 = 1$.

When repeated factors have the same base, they can be written together as a power of that base. For example, $7 \cdot 7 \cdot 7$ can be written as 7^3. When letters are used instead of numbers, we can't compute the basic numeral, but we can still combine repeated factors together as a power if they have the same base. For example, we may write $b \cdot b \cdot b \cdot b$ as b^4.

Example 3   Use Exponents for Repeated Factors

Write each expression using exponents.

(a) $6 \cdot 6 \cdot 6 \cdot 8 \cdot 8$ (b) $b \cdot c \cdot b \cdot c \cdot c$

(a) We group the three factors of 6 together as 6^3 and the two factors of 8 together as 8^2. So

$$6 \cdot 6 \cdot 6 \cdot 8 \cdot 8 = 6^3 \cdot 8^2$$

(b) $b \cdot c \cdot b \cdot c \cdot c = b \cdot b \cdot c \cdot c \cdot c$ Commutative Property

$\qquad\qquad\quad = (b \cdot b)(c \cdot c \cdot c)$ Associative Property

$\qquad\qquad\quad = b^2 \cdot c^3$ Definition of exponents

EXERCISE 6.1

DEVELOP YOUR SKILL

Use the definition of a power to construct an operator model with repeated multipliers. Then compute.

1. 2^4 **2.** 4^2 **3.** 5^2 **4.** 2^5

Compute the basic numeral.

5. 6^2 **6.** 7^2 **7.** 5^0 **8.** 5^3

9. 3^4 **10.** 7^0 **11.** 8^2 **12.** 9^2

Write each expression using exponents.

13. $5 \cdot 5 \cdot 5 \cdot 5$ **14.** $3 \cdot 7 \cdot 7 \cdot 3 \cdot 7$

15. $d \cdot d \cdot d$ **16.** $x \cdot y \cdot x \cdot y \cdot x$

MAINTAIN YOUR SKILL

17. Diane had 128 stamps in her drawer. After mailing her Christmas cards, she only had 43 stamps left. Let x be the number of stamps she used on the cards. Write a subtraction equation with x as the variable to model this problem. Then solve the equation to find how many stamps she used on the cards. [5.7]

Compute the basic form for the composite of the two operators. [2.4, 4.2]

18. $-8 + 5$ **19.** $-3 + 7$

20. $8\left(\dfrac{(\ )}{4}\right)$ **21.** $3\left(\dfrac{(\ )}{12}\right)$

Each expression is an operand followed by two operators. Do not compute and do not change the operand, but use parentheses to write an equivalent expression with the composite operator. When possible, give an alternate form. [2.5]

22. $36 - 21 + 27$ **23.** $34 - 18 - 5$

24. $25 + 6 - 14$ **25.** $20 + 15 - 8$

6.2 – EXPONENTS AND THE ORDER OF OPERATIONS

OBJECTIVE

1. Use the order of operations with expressions involving exponents.

The diagram in Figure 6.1 shows the six basic operations used in arithmetic and algebra. The words in quotes are the names of the forms. The three operations on the left side (addition, multiplication, and exponentiation) are all *direct* operations. Addition can be defined in terms of counting. Multiplication is repeated addition and exponentiation is repeated multiplication.

The three operations on the right side are *indirect* operations. They are defined by their inverse relation to the direct operation to their left. Subtraction is the inverse of addition. Division is the inverse of multiplication. The inverse of exponentiation is called "taking a root," but it will not be covered in this course. The inverse operations are only partial operations on the counting numbers. That is, they may not be defined for some choices of counting numbers.

Figure 6.1

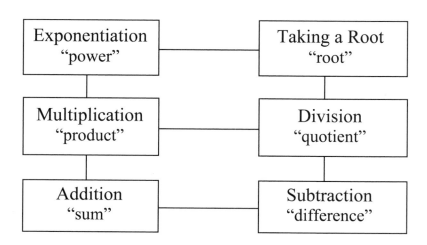

The direct operations of addition and multiplication are both commutative and associative, but the indirect operations of subtraction and division have neither property. The direct operation of raising to a power is *not* commutative or associative, as shown by the counterexamples below.

Example 1   Powers are Not Commutative

What happens if we interchange the exponent and the base in a power? Does the value remain the same? The answer is no, not necessarily. For example,

$$2^3 = 1 \cdot 2 \cdot 2 \cdot 2 = 8,$$

but $3^2 = 1 \cdot 3 \cdot 3 = 9$. So $2^3 \neq 3^2$.

In general, powers are not commutative: $x^y \neq y^x$.

Example 2   Powers are Not Associative

What happens if the exponent is itself a power? For example we have

$$2^{\left(3^2\right)} = 2^9 = 512.$$

If we regroup the terms so that the 3 is grouped with the base of 2, the value changes.

$$\left(2^3\right)^2 = 8^2 = 64. \text{ So } 2^{\left(3^2\right)} \neq \left(2^3\right)^2.$$

In general, powers are not associative: $x^{\left(y^z\right)} \neq \left(x^y\right)^z$.

We are now using five different operations from three levels on the chart in Figure 6.1. When more than one operation is used in an expression, the following rules tell us what order to follow when performing the operations.

Order of Operations

Unless there are instructions otherwise,

        raising to a power is done first,

        then multiplication and division,

        and finally, addition and subtraction.

Briefly, start high on Figure 6.1 and work down. With operations from the same level, work from left to right.

Parentheses and other grouping symbols, such as the bar of division, are used to make exceptions to these rules. We always do what is inside the parentheses first. Remember that the name given to an entire expression comes from the *last* operation used in computing the basic numeral.

Study Tip:

The name of the form comes from the **last** operation used in computing the basic numeral.

Example 3   Name the Form

Name each form and compute the basic numeral.

   (a) $5 \cdot 2^3$ (b) $(5 \cdot 2)^3$ (c) $8 - 2 \cdot 3$

(a) We do exponents before multiplication, so this is a product. $5 \cdot 2^3 = 5 \cdot 8 = 40$.

(b) We work inside the parentheses first, so this is a power.

$$(5 \cdot 2)^3 = 10^3 = 1000.$$

(c) Multiplication comes before subtraction. This is a difference. $8 - 2 \cdot 3 = 8 - 6 = 2$.

When we have an expression using letters as variables, the value of the expression may change depending on the values that are given to each variable. In an expression like $x^2 + xy$, we can let x and y be any counting numbers. The value of $x^2 + xy$ is then found by substituting these values for x and y. For example, if $x = 3$ and $y = 5$, then

$$x^2 + xy = 3^2 + (3)(5) = 9 + 15 = 24.$$

Notice that in making the substitution, we replace x by 3 in both places that x appears.

Study Tip:

Note in part (b) that only the x gets squared, not the $2x$. This is because powers are computed before products.

Example 4   Evaluate an Expression

Evaluate each expression *when* $x = 5$ *and* $y = 7$.

   (a) $x^2 + xy$ (b) $2x^2 - 3y$

Substitute 5 for x and 7 for y and compute:

(a) $x^2 + xy = 5^2 + (5)(7) = 25 + 35 = 60$

(b) $2x^2 - 3y = 2(5^2) - 3(7)$

$$= (2)(25) - (3)(7) = 50 - 21 = 29$$

EXERCISE 6.2

DEVELOP YOUR SKILL

Name each form and compute the basic numeral.

1. $3^2 + 4^2$ 2. $(3 + 4)^2$ 3. $(6 - 3)^2$

4. $6^2 - 3^2$ 5. $2 \cdot 3^2$ 6. $(2 \cdot 3)^2$

7. $2 + 5^2$ 8. $(2 + 5)^2$ 9. $12 - 2^3$

Evaluate each expression when $x = 5$ and $y = 2$.

10. $x + 3y^2$

11. $x^0 y^3$

12. $x^2 - (x - y)^2$

MAINTAIN YOUR SKILL

Find the inverse of each operator. [6.2]

13. $+ 23$

14. $- 18$

15. $12(\ )$

16. $\dfrac{(\ )}{10}$

View each equation as consisting of exactly three parts. Identify the part that is the sum (S) or the product (P) of the other two parts. Give all possible interpretations. [5.7]

17. $4 + x = \dfrac{n}{7}$

18. $5c = 38 - x$

19. $y - 3 = \dfrac{n}{x}$

20. $\dfrac{x}{a} = y + 6$

Solve each equation for x. These are the same equations used in Exercises $17 - 20$. [5.7]

21. $4 + x = \dfrac{n}{7}$

22. $5c = 38 - x$

23. $y - 3 = \dfrac{n}{x}$

24. $\dfrac{x}{a} = y + 6$

6.3 – PRODUCTS AND QUOTIENTS OF POWERS

OBJECTIVE

1. View powers as operators and simplify products and quotients of powers.

Raising to a power may be viewed as an operation on the base. For example, 5^2 is the transform when the exponential operator $(\ )^2$ is joined to the operand 5.

$$\text{Operand: } 5 \quad \text{Operator: } (\ )^2 \quad \longrightarrow \quad \text{Transform: } 5^2$$

In general, the power b^2 is the transform when the exponential operator $(\ )^2$ is joined to the operand b. This power, b^2, may be read as "the second power of b," or as "the square of b." The power b^3 is "the third power of b" or "the cube of b," so that $(\ )^3$ is the cubing operator. In general, b^n is read as "the n-th power of b," or "b to the n."

Parentheses are a necessary part of exponential operators, $(\ )^n$, when they are written without an operand. They are also needed when the operand is a sum, difference, product, quotient, or power. Without the parentheses, an exponent refers only to the single number or variable at its lower left. For example,

$$2 + 3^2 = 2 + 9 = 11 \quad \text{and} \quad (2 + 3)^2 = 5^2 = 25.$$

The operator $(\ )^0$ cannot be used with zero as an operand since 0^0 is not defined. But $(\ )^0$ dominates every nonzero operand since the transform is always 1. For example, the definition of an exponent tells us to evaluate 5^0 by starting with 1 and multiplying by 5 no times. So, $5^0 = 1$.

Also, we see that $(\ )^1$ is the **identity** operator for raising to a power. For example, $5^1 = 1 \cdot 5 = 5$ and $7^1 = 1 \cdot 7 = 7$. Since this is true for any base b, we usually write b^1 more simply just as b.

What happens when we have a product or two powers with the same base? Can they be combined? The answer is yes.

Study Tip:

To multiply powers with the same base, add their exponents.

Example 1   A Product of Powers

$$
\begin{aligned}
7^3 \cdot 7^2 &= (7 \cdot 7 \cdot 7)(7 \cdot 7) && \text{definition} \\
&= 7 \cdot 7 \cdot 7 \cdot 7 \cdot 7 && \text{associative principle} \\
&= 7^5 && \text{definition}
\end{aligned}
$$

We see that the *product of two powers* that have the same base is equivalent to a *power* of that base. The exponent for the power is the *sum* of the exponents for the powers that are factors of the product.

Here is the general rule:

Product of Powers

$$x^m \cdot x^n = x^{m+n}$$

Something similar happens with quotients of powers.

Study Tip:

To divide powers with the same base, subtract their exponents.

Example 2   A Quotient of Powers

$$\frac{7^5}{7^2} = \frac{7 \cdot 7 \cdot 7 \cdot 7 \cdot 7}{7 \cdot 7} \qquad \text{definition}$$

$$= \frac{((7 \cdot 7 \cdot 7)(7 \cdot 7)}{(7 \cdot 7)} \qquad \text{associative principle}$$

$$= 7 \cdot 7 \cdot 7 \qquad \begin{array}{l}\text{Cancel the multiplying} \\ \text{and dividing by } (7 \cdot 7).\end{array}$$

$$= 7^3 \qquad \text{definition}$$

We see that a *quotient of two powers* with the same base is equivalent to a *power* of that base. The exponent for the quotient comes from the exponent in the dividend minus the exponent in the divisor.

Here is the general rule:

Quotient of Powers

$$\frac{x^m}{x^n} = x^{m-n}$$

Example 3   **Change a Product or a Quotient into a Power**

Change each product or quotient into an equivalent power. Do not compute, except as needed for exponents.

(a) $2^4 \cdot 2^3$ (b) $\dfrac{3^5}{3}$ (c) $9^4 \cdot 9^0$

(a) The bases are the same, so add the exponents. The product $2^4 \cdot 2^3$ is equal to the power $2^{4+3} = 2^7$.

(b) The quotient $\dfrac{3^5}{3}$ is equal to $\dfrac{3^5}{3^1}$, since $3 = 3^1$. Subtract the exponents: $3^{5-1} = 3^4$.

(c) As in part (a), we add the exponents. The product $9^4 \cdot 9^0$ is equal to the power $9^{4+0} = 9^4$. Recall that $9^0 = 1$, so multiplying by 9^0 is the same as multiplying by 1.

EXERCISE 6.3

DEVELOP YOUR SKILL

Change each product or quotient into an equivalent power. Do not compute, except for exponents.

1. $5^2 \cdot 5^4$ **2.** $3^2 \cdot 3^0 \cdot 3^3$ **3.** $\dfrac{8^6}{8^2}$

4. $\dfrac{7^{12}}{7^3}$

5. $\dfrac{4^5 \cdot 4^4}{4^3}$

6. $\dfrac{5^8}{5^4 \cdot 5^0}$

7. $\dfrac{5 \cdot 5^4}{5^2}$

8. $\dfrac{6^3 \cdot 6^0}{6}$

MAINTAIN YOUR SKILL

9. There were 24 orchids on the front of the float, 16 orchids on the sides of the float, and some more orchids on the back. If there were 58 orchids on the float, how many were on the back? [2.2]

In Exercises 10 – 11, an operand number is given, followed by operators to be used in the order stated. Build this form, but do not compute and do not use the commutative principle. Then name the form that has been built. [4.4]

10. $16; \; -4, \; \dfrac{(\;)}{4}, \; 5(\;)$

11. $7; \; (\;)2, \; +1, \; \dfrac{(\;)}{3}$

Analyze each expression by stating an operand and then listing a sequence of operators that can be joined one at a time to produce the given expression. If more than one operand and sequence of operators can be used, give the alternate possibilities. [4.5]

12. $\dfrac{5 \cdot 4 - 8}{3}$

13. $4\left(\dfrac{15}{3} + 3\right)$

Simplify each expression by canceling inverse operators. If no simpler form is possible, write "not possible." [4.6]

14. $\dfrac{x+2}{x-2}$

15. $\dfrac{4n+3}{4}$

16. $\dfrac{5(y-3)}{5}$

17. $\dfrac{x(n+1)}{n+1}$

18. $\dfrac{2n-3}{n-3}$

19. $\dfrac{5x+4}{5x+2}$

Name each form and compute the basic numeral. [6.3]

20. $2^2 + 5^2$

21. $(2+5)^2$

22. $(8-2)^2$

23. $8 - 2^2$

24. $3 \cdot 4^2$

25. $(3 \cdot 4)^2$

6.4 – PRIME FACTORED FORM

OBJECTIVE

1. Write the prime factorizations of composite numbers.

Study Tip:

A natural number greater than one is **prime** if its only factors are one and itself. Otherwise, it is **composite**.

The numbers 0 and 1 are neither prime nor composite.

The natural numbers greater than one can always be written as a product of two factors. For example, the number 15 can be written as a product in two ways:

$$1 \times 15 = 15 \quad \text{and} \quad 3 \times 5 = 15.$$

We see that 15 has four factors: 1, 3, 5, and 15. Likewise, the number 9 can be factored as

$$1 \times 9 = 9 \quad \text{and} \quad 3 \times 3 = 9,$$

so the number 9 has three factors: 1, 3, and 9.

The number 7, on the other hand can only be written as a product by multiplying 1 and 7:

$$1 \times 7 = 7.$$

The only factors of 7 are 1 and 7.

The numbers 15 and 9 are examples of **composite** numbers: they can be written as a product of two natural numbers each greater than 1. The number 7 is an example of a **prime** number: it cannot be written as a product of natural numbers each greater than 1. The numbers 0 and 1 are neither prime nor composite.

Every natural number greater than 1 is either prime or it can be written as a product of primes. If we decide on the order of writing the prime factors, then we have just one way of writing each number, called the **prime factored form**. This is shown in Example 1 for the numbers 2 through 21.

Example 1   Prime Factored Form

Basic Numeral	Prime Factored Form	Basic Numeral	Prime Factored Form	Basic Numeral	Prime Factored Form
		8	2^3	15	$3 \cdot 5$
2	2	9	3^2	16	2^4
3	3	10	$2 \cdot 5$	17	17
4	2^2	11	11	18	$2 \cdot 3^2$
5	5	12	$2^2 \cdot 3$	19	19
6	$2 \cdot 3$	13	13	20	$2^2 \cdot 5$
7	7	14	$2 \cdot 7$	21	$3 \cdot 7$

The following rules are used to standardize the writing of the prime factored form:

Rules for Prime Factored Form

Rule	Example
1. A prime is written as a prime.	1. $3 = 3$
2. A power of a prime is written as a power of a prime.	2. $8 = 2^3$
3. If there are two or more prime factors, the prime factors are arranged in order of increasing size and appear only once, using exponents as needed.	3. $18 = 2 \cdot 3^2$

Example 2   Prime Factored Form

In what way does each of the following fail to be the prime factored form for 36?

(a) $2^2 \cdot 9$ (b) $(2 \cdot 3)^2$ (c) $3^2 \cdot 2^2$ (d) $2 \cdot 2 \cdot 3^2$

(a) 9 is not a prime.

(b) This is a power of a product, not a product of powers.

(c) The prime 2 should be to the left of prime 3.

(d) It should be 2^2 instead of $2 \cdot 2$.

When the basic numeral is known for a number, its prime factored form can be found by repeatedly factoring until all the factors are prime. Like primes are grouped together using powers. Then the terms are arranged so that the factors (and the base of the powers) increase in size from left to right.

Study Tip:

In part (a), be sure to write $2^3 \cdot 3^2$ and not $3^2 \cdot 2^3$. The prime bases increase from left to right.

Example 3   Prime Factored Form

Find the prime factored form for each number.

(a) 72   (b) 200

(a) $72 = 6 \cdot 12 = (2 \cdot 3)(3 \cdot 4)$
$$= (2 \cdot 3)[3 \cdot (2 \cdot 2)] = 2^3 \cdot 3^2$$

(b) $200 = 2 \cdot 100 = 2 \cdot (10 \cdot 10)$
$$= 2 \cdot [(2 \cdot 5)(2 \cdot 5)] = 2^3 \cdot 5^2$$

Note that it does not matter how we begin the factoring. The final form will be the same. For example, we could have factored 200 as $20 \cdot 10$:

$$200 = 20 \cdot 10 = (4 \cdot 5)(2 \cdot 5)$$
$$= [(2 \cdot 2) \cdot 5](2 \cdot 5) = 2^3 \cdot 5^2$$

Example 4 **Prime Factored Form**

Find the prime factored form for each number.

(a) 420   (b) 53

(a) $420 = 42 \cdot 10 = (2 \cdot 21)(2 \cdot 5)$
$= [2 \cdot (3 \cdot 7)](2 \cdot 5) = 2^2 \cdot 3 \cdot 5 \cdot 7$

(b) Trials will show that none of the primes 2, 3, 5, or 7 will divide 53. The next prime to try is 11, but $11 \cdot 11 = 121$ and $121 > 53$. So if 53 were a product of two numbers, then one of these would have to be less than 11. Since trials have shown that 53 does not have a prime divisor less than 11, then 53 must itself be a prime. So the prime factored form is 53.

EXERCISE 6.4

DEVELOP YOUR SKILL

Write each number in prime factored form.

1. 21	**2.** 22	**3.** 23
4. 24	**5.** 25	**6.** 26
7. 27	**8.** 28	**9.** 29
10. 30	**11.** 32	**12.** 34

MAINTAIN YOUR SKILL

Solve each equation for the number 27, but do not compute. [5.1, 5.3]

13. $65 = 38 + 27$

14. $27 - 11 = 16$

15. $60 - 27 = 33$

16. $108 = 4 \cdot 27$

17. $\dfrac{27}{9} = 3$

18. $\dfrac{135}{27} = 5$

Solve each equation for x. [5.7]

19. $3y - 5 = \dfrac{a}{x}$

20. $5x = 2n + 1$

21. $5n - x = 8$

22. $x - 12 = \dfrac{b}{3}$

23. $\dfrac{x}{y} = n + 6$

24. $3y + x = 2k$

6.5 – COMMON FACTORS AND COMMON MULTIPLES

OBJECTIVE

1. Find the greatest common factor and the least common multiple of two or more numbers.

Since $25 = 5^2$ and $35 = 5 \cdot 7$, the numbers 25 and 35 have the factor 5 in common. We say that 5 is a **common factor** of 25 and 35. The largest of the common factors of two or more numbers is called the **greatest common factor (GCF)**. One way to find the GCF of two numbers is to list all their factors and see what is the largest factor in common. For example, for 25 and 35 we have

Factors of 25: 1, 5, 25
Factors of 35: 1, 5, 7, 35   } The GCF is 5.

This works well for small numbers, but with larger numbers it can be quite tedious. A better way is to examine the prime factored form for each number. The common factors show up clearly if the prime factors of the numbers are arranged in rows and columns as in Example 1. Here we have spread out the prime factors so that each column has at most one prime factor, and within each column all the factors are the same. The factors of the GCF come from the columns that are full—i.e., that contain a factor for each of the original numbers.

Example 1   Find the Greatest Common Factor

Find the GCF *of 210, 196, and 294.*

Make a chart that shows the prime factors.

$210 = 2 \cdot 3 \cdot 5 \cdot 7$	2		3	5	7	
$196 = 2^2 \cdot 7^2$	2	2			7	7
$294 = 2 \cdot 3 \cdot 7^2$	2		3		7	7
	2				7	← GCF

The first "2" column and the first "7" column are full, meaning that 2 and 7 are factors of all three numbers. We have placed a 2 and a 7 at the bottom of their respective full columns (below the solid line). The GCF comes from their product: $2 \cdot 7$ or 14. This is the largest number that is a factor of all three original numbers.

Just as two numbers will have at least one factor in common, they will also have multiples in common. The least of the *nonzero* common multiples is called the **least common multiple (LCM)**. For example, by listing the nonzero multiples of 25 and 35 we see that the LCM of 25 and 35 is 175:

Nonzero multiples of 25:

25, 50, 75, 100, 125, 150, **175**, 200, …

Nonzero multiples of 35:

35, 70, 105, 140, **175**, 210, 245, …

Once again, the process is simplified by using the prime factored form for each number. The next example is a repeat of Example 1 with their LCM included. We see that the factors of the LCM come from each of the columns.

Example 2   Find the Least Common Multiple

Find the LCM *of* 210, 196, *and* 294.

Repeat the chart from Example 1 and label each column at the top with its prime number. The LCM comes from taking all these labels.

	2	2	3	5	7	7	← LCM
$210 = 2 \cdot 3 \cdot 5 \cdot 7$	2		3	5	7		
$196 = 2^2 \cdot 7^2$	2	2			7	7	
$294 = 2 \cdot 3 \cdot 7^2$	2		3		7	7	
	2				7		← GCF

The LCM is $2 \cdot 2 \cdot 3 \cdot 5 \cdot 7 \cdot 7$, or $2^2 \cdot 3 \cdot 5 \cdot 7^2$, or 2940.

Example 3   Find the GCF and LCM

Find the GCF *and* LCM *of* 8 *and* 45.

Make a chart that shows the prime factors.

	2	2	2	3	3	5	← LCM
$8 = 2^3$	2	2	2				
$45 = 3^2 \cdot 5$				3	3	5	
							← GCF

This time there are no prime factors in the bottom row. This means the Greatest Common Factor is 1. The Least Common Multiple of 8 and 45 is their product, $2^3 \cdot 3^2 \cdot 5$.

Study Tip:

When the GCF of two numbers is 1, the numbers are called **relatively prime**. This is because they have no prime divisors in common.

EXERCISE 6.5

DEVELOP YOUR SKILL

Find the GCF and the LCM of each set of numbers. Write your answers in prime factored form.

1. $\begin{cases} 3 \cdot 5^2 \\ 2 \cdot 3^2 \end{cases}$

2. $\begin{cases} 2 \cdot 5 \cdot 7 \\ 2 \cdot 3 \cdot 5 \end{cases}$

3. $\begin{cases} 2 \cdot 3^2 \cdot 7 \\ 2^2 \cdot 7 \end{cases}$

4. $\begin{cases} 2 \cdot 5^2 \\ 3^2 \cdot 7 \end{cases}$

5. $\begin{cases} 2 \cdot 5^2 \\ 2^2 \cdot 5 \\ 3 \cdot 5 \end{cases}$

6. $\begin{cases} 2 \cdot 3 \\ 3 \cdot 5 \\ 2 \cdot 3^2 \end{cases}$

7. $\begin{cases} 2 \cdot 3^3 \cdot 5 \\ 2^2 \cdot 3^2 \\ 2 \cdot 3^4 \cdot 7 \end{cases}$

8. $\begin{cases} 2 \cdot 3 \\ 3^2 \cdot 5 \\ 2 \cdot 3 \cdot 5^2 \end{cases}$

9. Three separate strings of lights are used on a Christmas tree. Each has a blinker device. One string flashes off and on every 24 seconds, another every 30 seconds, and another every 36 seconds. If they are all seen to blink together at one time, how long will it be until this happens again?

MAINTAIN YOUR SKILL

10. Roger had one hundred twenty-seven baseball cards. He bought some more cards and then had one hundred fifty-three cards. Let x be the number of cards Roger bought. Write an addition equation with x as the variable to model this problem. Then solve the equation to find how many cards Roger bought. [5.7]

Each expression is an operand followed by two operators. Do not change the operand, but use parentheses to write an equivalent expression with the composite operator. Assume that $a \geq b$ and that all differences are defined. When possible, give an alternate form. [2.5]

11. $25 - a - b$

12. $23 - b + a$

13. $32 + a - b$

14. $30 + b - a$

Name each form and compute the basic numeral. [6.2]

15. $4 \cdot 5^2$

16. $3 \cdot 5 + 4$

17. $\dfrac{(2 \cdot 5)^2}{20}$

18. $\left(\dfrac{12}{2} + 1 \right)^2$

19. $18 - 5 \cdot 2$

Evaluate each expression when $x = 5$ and $y = 2$. [6.2]

20. $2y^0 + x^2$

21. $3xy - y^2$

22. $2x^2 - 4y$

Change each product or quotient into an equivalent power. Do not compute, except for exponents. [6.3]

23. $5^5 \cdot 5^2 \cdot 5^0$

24. $\dfrac{7^8}{7^2}$

25. $\dfrac{6^2 \cdot 6^6}{6^3}$

CHAPTER 6 REVIEW

Write each expression using exponents. [6.1]

1. $7 \cdot 7 \cdot 7$

2. $3 \cdot 7 \cdot 7 \cdot 3 \cdot 7$

Evaluate each expression when $x = 2$ and $y = 3$. [6.2]

3. $2x^2 + y^0$

4. $xy^2 - 5x$

Name each form and compute the basic numeral. [6.2]

5. $2 + 4^2$

6. $5^2 - 2^2$

7. $4(9 - 7)^2$

8. $2 \cdot 9 - 2^3$

9. $3 + 4 \cdot 5$

10. $\dfrac{8 + 7}{3}$

11. $\left(\dfrac{9 - 3}{2}\right)^2$

12. $3 \cdot 4^2$

Change each product or quotient into an equivalent power. Do not compute, except for exponents. [6.3]

13. $3^4 \cdot 3 \cdot 3^2$

14. $\dfrac{7^6}{7^2}$

15. $\dfrac{5^4 \cdot 5^3}{5^2}$

16. $\dfrac{3^0 \cdot 3^4}{3^2}$

Write each number in prime factored form. [6.4]

17. 36

18. 40

19. 45

20. 47

Find the GCF and the LCM of each set of numbers. Write your answers in prime factored form. [6.5]

21. $\begin{cases} 2 \cdot 3 \\ 3 \cdot 7 \end{cases}$

22. $\begin{cases} 3^2 \cdot 5^2 \\ 2 \cdot 3^2 \end{cases}$

23. $\begin{cases} 2^2 \cdot 7 \\ 3 \cdot 7 \\ 2 \cdot 3^2 \end{cases}$

24. $\begin{cases} 2^2 \cdot 3 \cdot 7 \\ 2 \cdot 5 \cdot 7^2 \\ 2 \cdot 3^2 \cdot 5 \end{cases}$

Chapter 7 — Rational Numbers

7.1 – QUOTIENTS AND FRACTIONS

OBJECTVES

1. Understand the meaning of a fraction and represent it graphically.

2. Reduce a fraction to lowest terms.

Up to this point we have only been using counting numbers. But sometimes we need more than just the counting numbers. For example, the quotient $\frac{2}{3}$ cannot be a counting number, since there is no counting number x such that $3x = 2$, as required by the definition of division. But the *need* for such a number can be seen in the following problem.

Suppose we have a board that is two units long:

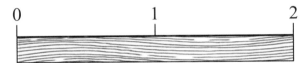

If we cut this board into 3 equal parts, how long will each part be?

The symbols that we use to describe the length above are the same symbols that were used to state the problem. Namely, when 2 is divided by 3, the result is $\frac{2}{3}$. So what is $\frac{2}{3}$? It is the number that when multiplied by 3 is equal to 2.

Definition of a Rational Number

The numbers x that are solutions of the equation

$$(n)(x) = m,$$

where m and n are counting numbers with $n \neq 0$, are called **rational numbers**. The number x can be represented by the **fraction** $\frac{m}{n}$. That is,

$$(n)\left(\frac{m}{n}\right) = m.$$

There are two helpful ways to visualize the fraction $\frac{2}{3}$. In the introduction we started with 1, multiplied by 2 (to get a board of length 2), and then we divided by 3. That is,

$$\frac{2}{3} = \frac{2 \cdot 1}{3}.$$

We can model this as follows:

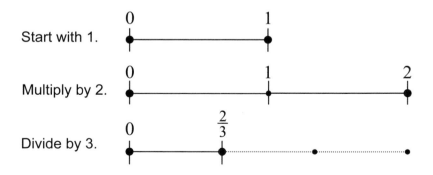

But we can also think of $\frac{2}{3}$ as starting with 1, dividing by 3 (to get $\frac{1}{3}$), and then multiplying by 2. That is, $\frac{2}{3} = (2)\left(\frac{1}{3}\right)$.

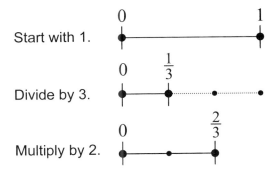

Notice that the length obtained here for $\frac{2}{3}$ is the same as the length obtained by taking 2 and dividing by 3.

In Lesson 4.2 we saw that multiplying by a and dividing by b, in either order, is the same as multiplying by the quotient $\frac{a}{b}$ when b divides a. We now extend this to the case where $\frac{a}{b}$ is a fraction and b does not necessarily divide a.

Study Tip:

This is Case 3 in Lesson 4.2.

$$\frac{a}{b}(\ ) = \frac{a(\ )}{b} = a\left(\frac{(\ )}{b}\right).$$

This means that the effect of multiplying by $\frac{2}{3}$ is the same as multiplying by 2 and dividing by 3, in either order. Since the contraction is larger than the expansion, the overall effect of multiplying by $\frac{2}{3}$ is to make a nonzero operand smaller.

In general, the effect of multiplying by a fraction is the combination of two actions—one an expansion and the other a contraction.

Example 1   **Effect of Multiplying by a Fraction**

Describe the effect of each operator as the composition of an expansion and a contraction. Then indicate whether the overall effect is to make a nonzero operand larger or smaller.

(a) $\frac{3}{4}(\ \ )$ (b) $\frac{5}{8}(\ \ )$ (c) $\frac{11}{3}(\ \ )$

(a) This is an expansion by 3 and a contraction by 4. Since the contraction is larger than the expansion, the overall effect is to make a nonzero operand smaller.

(b) This is an expansion by 5 and a contraction by 8. Since the contraction is larger than the expansion, the overall effect is to make a nonzero operand smaller.

(c) This is an expansion by 11 and a contraction by 3. This time the expansion is larger, so the overall effect is to make a nonzero operand larger.

Study Tip:

Sometimes multiplying by a fraction makes things larger. This is the case in part (c).

When $\frac{2}{3}$ is viewed as a quotient, we have called 2 the dividend and 3 the divisor. When $\frac{2}{3}$ is viewed as a fraction, then 2 is called the **numerator** and 3 is called the **denominator**.

Quotient	Fraction
dividend $\longrightarrow$ $\dfrac{2}{3}$ $\longleftarrow$ divisor	$\dfrac{2}{3}$ $\longleftarrow$ numerator, $\longleftarrow$ denominator

These new names come from viewing $\frac{2}{3}$ as $2\left(\frac{1}{3}\right)$. The denominator identifies (or denominates) the kind of a fraction—

in this case thirds. The numerator counts (or enumerates) how many of the thirds we have—in this case two.

If counting numbers are used for m and n, then $\frac{m}{n}$ is a **simple** fraction. If $m < n$, then $\frac{m}{n}$ is a **proper** fraction, but if $m \geq n$, then $\frac{m}{n}$ is called an **improper** fraction.

Example 2 **Simple Fractions and Proper Fractions**

$\frac{2}{3}$, $\frac{4}{8}$, and $\frac{7}{12}$ are simple proper fractions.

$\frac{5}{3}$, $\frac{10}{2}$, and $\frac{7}{7}$ are simple improper fractions.

$\frac{2 \cdot 3}{4}$, $\frac{\frac{2}{3}}{4}$, and $\frac{5}{3 + 8}$ are not simple fractions.

Every rational number has many fraction names by which it can be represented. For example,

$$\frac{2}{3} = \frac{4}{6} = \frac{6}{9} = \frac{8}{12} = \cdots$$

These equalities all follow from the invariant principle for quotients:

$$\frac{2}{3} = \frac{2 \cdot 2}{3 \cdot 2} = \frac{4}{6}, \quad \frac{2}{3} = \frac{2 \cdot 3}{3 \cdot 3} = \frac{6}{9}, \quad \frac{2}{3} = \frac{2 \cdot 4}{3 \cdot 4} = \frac{8}{12}, \cdots$$

Among all the equivalent fractions that represent the same rational number $\frac{x}{y}$, the **basic numeral** is the simple fraction with x and y as small as possible. Such a fraction is said to be in **lowest terms**. In reducing a fraction to lowest terms, the invariant principle for quotients is very useful. Recall that

$$\frac{x}{y} = \frac{nx}{ny} \quad \text{and} \quad \frac{x}{y} = \frac{x \div n}{y \div n}, \quad \text{when } n \neq 0.$$

To reduce a fraction to lowest terms, we look for any natural number that is a factor of both the numerator and the denominator. Then we can remove the same operator ( )n from the numerator and the denominator. This has the effect of dividing both the numerator and the denominator by n. Or, we may think of canceling a multiplying by n (in the numerator) with a dividing by n (in the denominator).

Study Tip:

Removing the · 7 operators has the effect of dividing both the numerator and the denominator by 7.

Example 3   Reduce a Fraction

Reduce the fraction $\frac{21}{28}$ to lowest terms.

$$\frac{21}{28} = \frac{3 \cdot 7}{4 \cdot 7}$$
Factor the numerator and the denominator.

$$= \frac{3 \cdot \cancel{7}}{4 \cdot \cancel{7}} = \frac{3}{4}$$
Remove the ( )7 operators. Or, cancel a multiplying by 7 (in the numerator) with a dividing by 7 (in the denominator).

The process of reducing a fraction can be accomplished in just one step if we find the GCF (greatest common factor) of the numerator and the denominator. For example,

$$\frac{30}{42} = \frac{5 \cdot 6}{7 \cdot 6} = \frac{5}{7}$$

But if we don't see the greatest common factor, the process can be done in repeated steps:

$$\frac{30}{42} = \frac{15 \cdot 2}{21 \cdot 2} = \frac{15}{21} = \frac{5 \cdot 3}{7 \cdot 3} = \frac{5}{7}$$

When the denominator of a fraction is 1, the fraction can be reduced to a counting number.

Example 4   **Reduce a Fraction**

Reduce the fraction $\frac{30}{15}$ *to lowest terms.*

$$\frac{30}{15} = \frac{2 \cdot 15}{1 \cdot 15} = \frac{2}{1} = 2$$

Notice how 15 was replaced by $1 \cdot 15$ so that 15 is seen as a factor in the denominator. When both the numerator and the denominator are divided by 15, we obtain 2 divided by 1. But any number divided by 1 is that same number.

EXERCISE 7.1

DEVELOP YOUR SKILL

Describe the effect of each operator as the composition of an expansion and a contraction. Then indicate whether the overall effect is to make a nonzero operand larger or smaller.

1. $\frac{4}{5}(\ )$

2. $\frac{3}{7}(\ )$

3. $\frac{5}{3}(\quad)$ **4.** $\frac{7}{4}(\quad)$

5. $\frac{2}{9}(\quad)$ **6.** $\frac{6}{5}(\quad)$

Indicate which of the following fractions are proper fractions.

7. $\frac{3}{6}, \frac{4}{3}, \frac{7}{8}$ **8.** $\frac{4}{6}, \frac{3}{7}, \frac{6}{5}$

Reduce each fraction to lowest terms.

9. $\frac{6}{8}$ **10.** $\frac{6}{9}$ **11.** $\frac{25}{35}$

12. $\frac{28}{40}$ **13.** $\frac{21}{3}$ **14.** $\frac{18}{24}$

15. $\frac{16}{32}$ **16.** $\frac{32}{8}$

MAINTAIN YOUR SKILL

17. The product of 5 and 8 is how much more than the sum of 5 and 8? [3.1]

For each equation, write the corresponding equation that is matched to it by the definition of subtraction or the definition of division. [1.5, 3.2]

18. $17 + x = y$ **19.** $21 - a = b$

20. $h \div k = 5$ **21.** $3m = n$

Commute the order of the two operators and then compute the basic form of the composite operator. [4.1, 4.2, 4.3]

22. $\frac{3(\quad)}{15}$ **23.** $28\left(\frac{(\quad)}{4}\right)$

24. $2\left(\frac{(\quad)}{16}\right)$ **25.** $\frac{24(\quad)}{6}$

7.2 – MULTIPLYING FRACTIONS

OBJECTIVES

1. Compute the product of two fractions.

2. Find the reciprocal of a rational number.

To compute the product of two fractions, we use the fact that multiplying by $\frac{m}{n}$ has the same effect as multiplying by m and dividing by n.

$$\left(\frac{m}{n}\right)\left(\frac{x}{y}\right) = \frac{m\left(\frac{x}{y}\right)}{n} \qquad \text{Multiply } \frac{x}{y} \text{ by } m \text{ and divide by } n.$$

$$= \frac{\frac{mx}{y}}{n} \qquad \text{Interchange the order of dividing by } y \text{ and multiplying by } m.$$

$$= \frac{mx}{ny} \qquad \text{The composition of dividing by } y \text{ and dividing by } n \text{ is dividing by } ny.$$

This gives us the following formula.

Product of Rational Numbers

> The product of the rational number $\frac{m}{n}$ and the rational number $\frac{x}{y}$ is the rational number $\frac{mx}{ny}$. That is,
>
> $$\left(\frac{m}{n}\right)\left(\frac{x}{y}\right) = \frac{mx}{ny}.$$

Example 1   **Multiply Fractions**

$$\left(\frac{2}{3}\right)\left(\frac{4}{5}\right) = \frac{2\cdot4}{3\cdot5} = \frac{8}{15} \qquad \left(\frac{3}{4}\right)\left(\frac{7}{8}\right) = \frac{3\cdot7}{4\cdot8} = \frac{21}{32}$$

Sometimes when two fractions are multiplied, the result is a basic numeral as above. Other times, the result is not a basic numeral and it must then be reduced to lowest terms. We can make this easier by writing all the numerators and denominators in factored form. Then we can divide the numerator and the denominator by any common factors to simplify the expression. This can be done after multiplying, as in part (a) of the next example, or before multiplying, as in part (b).

Example 2   Multiply Fractions

Compute the product and simplify: $\left(\dfrac{5}{6}\right)\left(\dfrac{4}{15}\right)$.

We factor the numerators and denominators. Then divide both by any common factors.

(a) $\left(\dfrac{5}{6}\right)\left(\dfrac{4}{15}\right) = \left(\dfrac{5}{2\cdot 3}\right)\left(\dfrac{2\cdot 2}{3\cdot 5}\right) = \dfrac{\cancel{5}\cdot\cancel{2}\cdot 2}{\cancel{2}\cdot 3\cdot 3\cdot\cancel{5}} = \dfrac{2}{9}$

(b) $\left(\dfrac{5}{6}\right)\left(\dfrac{4}{15}\right) = \left(\dfrac{1\cdot\cancel{5}}{\cancel{2}\cdot 3}\right)\left(\dfrac{\cancel{2}\cdot 2}{3\cdot\cancel{5}}\right) = \dfrac{2}{9}$

We could also have simplified the computation by dividing both 5 and 15 by 5 (their greatest common factor), and by dividing both 6 and 4 by 2, before multiplying numerators and denominators.

$$\left(\dfrac{\overset{1}{\cancel{5}}}{\underset{3}{\cancel{6}}}\right)\left(\dfrac{\overset{2}{\cancel{4}}}{\underset{3}{\cancel{15}}}\right) = \dfrac{2}{9}$$

A mixed number like $2\frac{1}{2}$ is really the sum of two numbers:

$$2\frac{1}{2} \quad \text{means} \quad 2 + \frac{1}{2}.$$

Because of this, we change mixed numbers into improper fractions before multiplying.

Example 3   Multiply Mixed Numbers

(a) $\left(2\frac{1}{2}\right)\left(3\frac{2}{3}\right) = \left(\frac{5}{2}\right)\left(\frac{11}{3}\right) = \frac{55}{6} \quad \text{or} \quad 9\frac{1}{6}$

(b) $(6)\left(4\frac{1}{8}\right) = (\overset{3}{\cancel{6}})\left(\frac{33}{\underset{4}{\cancel{8}}}\right) = \frac{99}{4} \quad \text{or} \quad 24\frac{3}{4}$

When the numerator and the denominator of a fraction change places, the new fraction and the original fraction are called **reciprocals** of each other. The product of a number and its reciprocal is always equal to 1.

For example, the reciprocal of $\frac{2}{3}$ is $\frac{3}{2}$ and

$$\left(\frac{2}{3}\right)\left(\frac{3}{2}\right) = \frac{6}{6} = 1.$$

In general, if a rational number is in the form of a simple fraction $\frac{x}{y}$, then its reciprocal is $\frac{y}{x}$ and $\left(\frac{x}{y}\right)\left(\frac{y}{x}\right) = 1$. The number zero

has a fraction or quotient form of $\frac{0}{1}$. But zero *does not* have a reciprocal, since the quotient $\frac{1}{0}$ is undefined.

Example 4   **Find a Reciprocal**

Find the reciprocal of each number.

(a) $\frac{3}{5}$ (b) $\frac{9}{4}$ (c) $\frac{1}{2}$ (d) 4   (e) $2\frac{1}{3}$

(a) The reciprocal of $\frac{3}{5}$ is $\frac{5}{3}$.

(b) The reciprocal of $\frac{9}{4}$ is $\frac{4}{9}$.

(c) The reciprocal of $\frac{1}{2}$ is $\frac{2}{1}$, or 2.

(d) The reciprocal of 4 (or $\frac{4}{1}$) is $\frac{1}{4}$.

(e) To find the reciprocal of $2\frac{1}{3}$, we first change it into the equivalent improper fraction $\frac{7}{3}$. The reciprocal of $\frac{7}{3}$ is $\frac{3}{7}$, and this is also the reciprocal of $2\frac{1}{3}$.

   Note that $2\frac{1}{3} \times \frac{3}{7} = \frac{7}{3} \times \frac{3}{7} = 1$, so $2\frac{1}{3}$ and $\frac{3}{7}$ are reciprocals.

EXERCISE 7.2

DEVELOP YOUR SKILL

Compute the basic numeral.

1. $\left(\dfrac{2}{3}\right)\left(\dfrac{3}{8}\right)$

2. $\left(\dfrac{3}{8}\right)\left(\dfrac{2}{5}\right)$

3. $\left(\dfrac{4}{3}\right)\left(\dfrac{2}{5}\right)$

4. $\left(\dfrac{3}{5}\right)\left(\dfrac{7}{6}\right)$

5. $\left(\dfrac{3}{4}\right)\left(\dfrac{8}{9}\right)$

6. $\left(\dfrac{15}{7}\right)\left(\dfrac{21}{5}\right)$

7. $\left(\dfrac{5}{6}\right)\left(\dfrac{2}{15}\right)$

8. $\left(\dfrac{4}{14}\right)\left(\dfrac{7}{6}\right)$

9. $\left(\dfrac{10}{12}\right)\left(\dfrac{8}{15}\right)$

10. $\left(\dfrac{6}{24}\right)\left(\dfrac{16}{20}\right)$

Compute the basic numeral. Use the mixed number form for your answer.

11. $\left(5\dfrac{1}{3}\right)\left(1\dfrac{1}{4}\right)$

12. $\left(2\dfrac{2}{5}\right)\left(2\dfrac{1}{3}\right)$

Write the reciprocal as a simple fraction or a natural number.

13. $\dfrac{3}{8}$

14. $4\dfrac{1}{2}$

15. 8

16. $\dfrac{1}{6}$

MAINTAIN YOUR SKILL

Use the definition of a power to construct an operator model with repeated multipliers. Then compute. [6.1]

17. 6^2

18. 4^3

Name each form and compute the basic numeral. [6.2]

19. $5 + 3^2$

20. $(7 - 4)^3$

21. $3(4 + 1)^2$

22. $\dfrac{14 + 10}{2^2}$

23. $12 - 3^2$

24. $\dfrac{8}{2} + 3 \times 5$

7.3 – DIVIDING FRACTIONS

OBJECTIVE

1. Compute
the quotient of
two fractions.

Since division with natural numbers is just a matter of writing the quotient, we would expect that division should be a simple operation with fractions. For example, to divide $\frac{2}{3}$ by $\frac{5}{7}$, we get the quotient just by writing

$$\frac{2}{3} \div \frac{5}{7} = \frac{\dfrac{2}{3}}{\dfrac{5}{7}}.$$

While the form at the right does represent the correct quotient number, it is not a basic numeral. In order to compute the basic numeral we use the invariant principle for quotients, which we now extend to rational numbers.

Invariant Principle for Quotients

If x, y, and n are rational numbers with $y \neq 0$ and $n \neq 0$, then

$$\frac{x}{y} = \frac{xn}{yn} \quad \text{and} \quad \frac{x}{y} = \frac{x \div n}{y \div n}.$$

To simplify the quotient $\frac{2}{3} \div \frac{5}{7}$, we seek an appropriate rational number n to use as a multiplier in the numerator and the denominator:

$$\frac{\dfrac{2}{3}}{\dfrac{5}{7}} = \frac{\left(\dfrac{2}{3}\right)n}{\left(\dfrac{5}{7}\right)n}$$

One way to do this is to let *n* be the reciprocal of the denominator. This gives us

$$\frac{\dfrac{2}{3}}{\dfrac{5}{7}} = \frac{\left(\dfrac{2}{3}\right)\left(\dfrac{7}{5}\right)}{\left(\dfrac{5}{7}\right)\left(\dfrac{7}{5}\right)} = \frac{\dfrac{14}{15}}{1} = \frac{14}{15}.$$

Since a number times its reciprocal is always 1, this process can be shortened to

$$\frac{\dfrac{2}{3}}{\dfrac{5}{7}} = \left(\dfrac{2}{3}\right)\left(\dfrac{7}{5}\right) = \frac{14}{15}.$$

That is, dividing by a fraction gives the same quotient as multiplying by its reciprocal. Instead of dividing by $\frac{5}{7}$, we can multiply by $\frac{7}{5}$:

$$\frac{2}{3} \div \frac{5}{7} = \frac{2}{3} \times \frac{7}{5} = \frac{14}{15}.$$

Here is the general rule.

Division of Fractions

> To divide by a fraction, we may multiply by its reciprocal:
> $$\frac{a}{b} \div \frac{c}{d} = \frac{a}{b} \times \frac{d}{c} = \frac{ad}{bc},$$
> where b, c, and d are not 0.

This rule should come as no surprise, given our work with expansions and contractions. For example,

$$\frac{2}{8}(\ ) = \frac{2(\ )}{8}$$

since multiplying by $\frac{2}{8}$ is the same as multiplying by 2 and dividing by 8. But

$$\frac{2(\ )}{8} = \frac{(\ )}{\frac{8}{2}}$$

Study Tip:

This is
Case 4 in
Lesson 4.2.

since the composition of an expansion by 2 and a contraction by 8 is a contraction by $\frac{8}{2}$. Combining these two equations we have

$$\frac{(\ )}{\frac{8}{2}} = \frac{2(\ )}{8} = \frac{2}{8}(\ )$$

Once again we see that dividing by a fraction has the same effect as multiplying by its reciprocal.

Example 1   **Divide Fractions**

Compute the basic numeral: $\frac{2}{5} \div \frac{3}{5}$.

Instead of dividing by $\frac{3}{5}$, we multiply by $\frac{5}{3}$.

$$\frac{2}{5} \div \frac{3}{5} = \left(\frac{2}{\cancel{5}}\right)\left(\frac{\cancel{5}^{\,1}}{3}\right) = \frac{2}{3}$$

Example 2   **Divide Mixed Numbers**

Compute the basic numeral: $1\dfrac{3}{4} \div 2\dfrac{1}{4}$.

We begin by writing each mixed number as an improper fraction.

$$\frac{1\dfrac{3}{4}}{2\dfrac{1}{4}} = \frac{\dfrac{7}{4}}{\dfrac{9}{4}} = \left(\frac{7}{\cancel{4}}\right)\left(\frac{\overset{1}{\cancel{4}}}{9}\right) = \frac{7}{9}$$

When solving an equation with fractions, we can use either method from Chapter 5.

Example 3   **Solve an Equation with Fractions**

Solve the equation $\dfrac{3}{4}x = \dfrac{2}{5}$ *for x.*

We will do this both ways.

(a) We can see this as a product and two factors, where x is a factor.

$$\left(\frac{3}{4}\right)(x) = \left(\frac{2}{5}\right)$$

factor   factor   product

So x is equal to the product $\frac{2}{5}$ divided by the other factor $\frac{3}{4}$.

$$x = \frac{2}{5} \div \frac{3}{4} = \frac{2}{5} \cdot \frac{4}{3} = \frac{8}{15}$$

(b) To get x by itself, we want to remove the multiplying by $\frac{3}{4}$ from the left side. When doing this we must join a dividing by $\frac{3}{4}$ to the right side.

$$\frac{3}{4}x = \frac{2}{5}$$

$$x = \frac{2}{5} \div \frac{3}{4} = \frac{2}{5} \cdot \frac{4}{3} = \frac{8}{15}$$

This second approach can be shortened somewhat if we realize that dividing by $\frac{3}{4}$ is the same as multiplying by its reciprocal $\frac{4}{3}$.

$$\frac{3}{4}x = \frac{2}{5}$$

$$\frac{4}{3}\left(\frac{3}{4}x\right) = \frac{2}{5} \cdot \frac{4}{3} \qquad \text{Multiply both sides by } \frac{4}{3}.$$

$$x = \frac{8}{15} \qquad \text{Simplify and compute.}$$

Study Tip:

With practice you can do the middle steps mentally and go directly from

$$\frac{3}{4}x = \frac{2}{5}$$

to

$$x = \frac{2}{5} \cdot \frac{4}{3} = \frac{8}{15}.$$

EXERCISE 7.3

DEVELOP YOUR SKILL

Compute the basic numeral.

1. $\dfrac{2}{5} \div \dfrac{4}{7}$

2. $\dfrac{3}{7} \div \dfrac{3}{5}$

3. $\dfrac{1}{2} \div \dfrac{5}{8}$

4. $\dfrac{\frac{3}{4}}{\frac{3}{8}}$

5. $\dfrac{\frac{4}{9}}{\frac{5}{9}}$

6. $\dfrac{\frac{3}{8}}{\frac{5}{6}}$

7. $\dfrac{\frac{5}{6}}{3\frac{1}{2}}$

8. $\dfrac{2\frac{2}{3}}{3\frac{1}{3}}$

Solve each equation for x.

9. $\dfrac{2}{3}x = \dfrac{1}{6}$

10. $\dfrac{4}{3}x = \dfrac{8}{7}$

11. $\dfrac{x}{\frac{3}{2}} = \dfrac{4}{5}$

12. $\dfrac{2}{3} = \dfrac{\frac{5}{6}}{x}$

13. $\dfrac{3}{5} = \dfrac{3}{4}x$

14. $\dfrac{2}{7} = \dfrac{4}{3}x$

MAINTAIN YOUR SKILL

Use distributive principles to replace each sum or difference by an equivalent product or quotient. Do <u>not</u> compute and do not use the commutative principle. [3.7]

15. $\dfrac{12}{4} + \dfrac{20}{4}$

16. $\dfrac{m}{3} - \dfrac{21}{3}$

17. $6x + 6y$

18. $7k - 4k$

Write a sentence that describes the action of the two operators. Then compute the basic form of the composite operator. [4.1, 4.2]

19. $30\left(\dfrac{(\ )}{2}\right)$

20. $\dfrac{27(\ )}{3}$

21. $\dfrac{\dfrac{(\ )}{12}}{3}$

Commute the order of the two operators. [4.3]

22. $30\left(\dfrac{(\ )}{2}\right)$

23. $\dfrac{27(\ )}{3}$

24. $\dfrac{\dfrac{(\ )}{12}}{3}$

7.4 – ADDING AND SUBTRACTING FRACTIONS

OBJECTIVE

1. Compute the sum or difference of fractions.

Study Tip:

When the denominators are the same, we just add the numerators.

When two fractions have the same denominator, they can be added using the distributive property of division.

Example 1   Add Like Fractions

Compute the basic numeral: $\dfrac{2}{7}+\dfrac{3}{7}$.

We use the distributive property, and then compute.

$$\frac{2}{7}+\frac{3}{7}=\frac{2+3}{7}=\frac{5}{7}$$

In order to add two fractions with different denominators, we first change them into equivalent fractions having a common denominator. Then we add the numerators.

Study Tip:

The Invariant Principle for Quotients tells us we may multiply the numerator and the denominator of a fraction by the same number and not change its value.

Example 2   Add Unlike Fractions

Compute the basic numeral: $\dfrac{1}{5} + \dfrac{2}{3}$

If we multiply the first fraction by $\dfrac{3}{3}$ and the second fraction by $\dfrac{5}{5}$, the equivalent fractions will both have 15 as a denominator.

$$\frac{1}{5} + \frac{2}{3} = \frac{1 \cdot 3}{5 \cdot 3} + \frac{2 \cdot 5}{3 \cdot 5} = \frac{3}{15} + \frac{10}{15}$$

Now we can use the distributive principle on the like fractions, and compute.

$$\frac{3}{15} + \frac{10}{15} = \frac{3 + 10}{15} = \frac{13}{15}$$

Since $\dfrac{a}{b} + \dfrac{c}{d} = \dfrac{ad}{bd} + \dfrac{bc}{bd} = \dfrac{ad + bc}{bd}$, we have the following general rule.

Addition of Fractions

> The sum of the rational numbers $\dfrac{a}{b}$ and $\dfrac{c}{d}$ may be computed as $\dfrac{a}{b} + \dfrac{c}{d} = \dfrac{ad + bc}{bd}$.

While this rule always produces a correct sum, it may not be a fraction in lowest terms. We can save work in reducing our answer if we use the least common multiple of b and d as the common denominator instead of their product. We refer to this number as the **Least Common Denominator** (LCD).

Example 3   **Add Unlike Fractions**

Compute the basic numeral: $\dfrac{1}{6} + \dfrac{7}{15}$.

We can find the sum by using the general rule:

$$\frac{1}{6} + \frac{7}{15} = \frac{1 \cdot 15 + 6 \cdot 7}{6 \cdot 15} = \frac{15 + 42}{90} = \frac{57}{90} = \frac{3 \cdot 19}{3 \cdot 30} = \frac{19}{30}.$$

But since the LCM (least common multiple) of 6 and 15 is 30, we can shorten our work and keep the numbers smaller by using 30 as the common denominator:

$$\frac{1}{6} + \frac{7}{15} = \frac{1 \cdot 5}{6 \cdot 5} + \frac{7 \cdot 2}{15 \cdot 2} = \frac{5}{30} + \frac{14}{30} = \frac{19}{30}$$

Study Tip:

The least common multiple of 6 and 15 is 30, so we use 30 as the LCD.

For the difference of two fractions to be defined, the first fraction must be greater than or equal to the second fraction. We have the following general rule:

Subtraction of Fractions

If $\dfrac{a}{b} \geq \dfrac{c}{d}$, then the difference $\dfrac{a}{b} - \dfrac{c}{d}$ is the rational

number $\dfrac{ad - bc}{bd}$.

As in addition, the computational work in subtraction can be shortened by using the least common multiple of b and d as the common denominator instead of their product.

Example 4   Subtract Unlike Fractions

Compute the basic numerals.

(a) $\dfrac{3}{8} - \dfrac{1}{6}$ (b) $\dfrac{7}{2} - \dfrac{5}{6}$ (c) $\dfrac{7}{6} - \dfrac{3}{4}$

(a) The LCD for 8 and 6 is 24, so multiply the first fraction by $\dfrac{3}{3}$ and the second fraction by $\dfrac{4}{4}$.

$$\frac{3}{8} - \frac{1}{6} = \frac{3\cdot 3}{8\cdot 3} - \frac{1\cdot 4}{6\cdot 4} = \frac{9}{24} - \frac{4}{24}$$

$$= \frac{9-4}{24} = \frac{5}{24}$$

Study Tip:

It may be necessary to do some reducing even when the LCD is used.

(b) The LCD is 6, so multiply the first fraction by $\dfrac{3}{3}$.

$$\frac{7}{2} - \frac{5}{6} = \frac{7\cdot 3}{2\cdot 3} - \frac{5}{6} = \frac{21}{6} - \frac{5}{6}$$

$$= \frac{21-5}{6} = \frac{16}{6} = \frac{8}{3}$$

(c) The LCD for 6 and 4 is 12, so multiply the first fraction by $\dfrac{2}{2}$ and the second fraction by $\dfrac{3}{3}$.

$$\frac{7}{6} - \frac{3}{4} = \frac{7\cdot 2}{6\cdot 2} - \frac{3\cdot 3}{4\cdot 3} = \frac{14}{12} - \frac{9}{12} = \frac{14-9}{12} = \frac{5}{12}$$

Example 5   **Add Mixed Numbers**

Compute the basic numeral: $2\frac{3}{4} + 3\frac{5}{6}$.

We can begin by changing the mixed numbers into improper fractions. We have $2\frac{3}{4} = \frac{11}{4}$ and $3\frac{5}{6} = \frac{23}{6}$. The LCD will be 12.

$$2\frac{3}{4} + 3\frac{5}{6} = \frac{11}{4} + \frac{23}{6} = \frac{11 \cdot 3}{4 \cdot 3} + \frac{23 \cdot 2}{6 \cdot 2}$$

$$= \frac{33}{12} + \frac{46}{12} = \frac{79}{12} = 6\frac{7}{12}$$

Or, we can combine the whole numbers and fractions separately.

$$2\frac{3}{4} + 3\frac{5}{6} = \left(2 + \frac{3}{4}\right) + \left(3 + \frac{5}{6}\right) \quad \text{Expand the mixed numbers.}$$

$$= (2 + 3) + \left(\frac{3}{4} + \frac{5}{6}\right) \quad \text{Use the commutative and associative properties.}$$

$$= 5 + \left(\frac{9}{12} + \frac{10}{12}\right) \quad \text{Get a common denominator.}$$

$$= 5 + \frac{19}{12} \quad \text{Add the numerators.}$$

$$= 5 + \left(1 + \frac{7}{12}\right) \quad \text{Replace } \frac{19}{12} \text{ by a mixed number.}$$

$$= 6\frac{7}{12} \quad \text{Add 5 + 1.}$$

The second way may appear to be longer, but it keeps the fractions smaller and some of the steps may be done mentally without writing them down.

Example 6   **Solve an Equation with Fractions**

Solve the equation $\dfrac{9}{8} - x = \dfrac{3}{4}$ *for x.*

$$\frac{9}{8} - x = \frac{3}{4}$$ Write the equation.

$$x = \frac{9}{8} - \frac{3}{4}$$ The addend x is the sum minus the other addend.

$$x = \frac{9}{8} - \frac{3 \cdot 2}{4 \cdot 2}$$ The LCD is 8. Multiply the second fraction by $\frac{2}{2}$.

$$x = \frac{9}{8} - \frac{6}{8}$$ Compute.

$$x = \frac{3}{8}$$ Combine by subtracting the numerators.

EXERCISE 7.4

DEVELOP YOUR SKILL

Compute the basic numeral.

1. $\dfrac{3}{5} + \dfrac{2}{3}$

2. $\dfrac{2}{3} - \dfrac{3}{7}$

3. $\dfrac{7}{10} - \dfrac{3}{5}$

4. $\dfrac{5}{6} - \dfrac{1}{3}$

5. $\dfrac{3}{10} + \dfrac{4}{15}$

6. $\dfrac{5}{12} + \dfrac{1}{8}$

7. $\dfrac{5}{6} - \dfrac{7}{15}$

8. $\dfrac{3}{5} + \dfrac{1}{15}$

9. $1\frac{1}{3} + 2\frac{1}{2}$ **10.** $5\frac{2}{3} + 6\frac{3}{4}$

Solve each equation for x.

11. $x + \frac{2}{3} = \frac{5}{4}$ **12.** $\frac{5}{2} - x = \frac{5}{3}$

13. $\frac{4}{3} = \frac{3}{5} + x$ **14.** $\frac{9}{8} = \frac{7}{4} - x$

15. $\frac{5}{6} - x = \frac{4}{15}$ **16.** $x + \frac{3}{10} = \frac{7}{4}$

MAINTAIN YOUR SKILL

17. Two hundred thirty-one pieces of chicken were evenly divided among the people at a family reunion and each person got three pieces. Let x be the number of people at the reunion. The total number of pieces of chicken divided by the number of people will give the number of pieces that each person received. Write a division equation with x as the variable to model this problem. Then solve the equation to find how many people attended the reunion. [5.7]

Each expression is an operand followed by two operators. Do not compute and do not change the operand, but use parentheses to write an equivalent expression with the composite operator. When possible, give an alternate form. [3.3, 3.5]

18. $x - 7 - 3$ **19.** $x - 12 + 6$

20. $x - 4 + 9$ **21.** $x + 5 - 8$

Compute the basic numeral. [7.2, 7.3]

22. $\left(\frac{3}{5}\right)\left(1\frac{1}{6}\right)$ **23.** $\left(\frac{3}{4}\right)\left(\frac{5}{12}\right)$

24. $\frac{5}{8} \div \frac{5}{3}$ **25.** $\frac{3}{5} \div \frac{11}{5}$

7.5 – ORDERING FRACTIONS

OBJECTIVES

1. Compare the size of two fractions.

2. Find fractions between two fractions.

If two fractions have the same denominator, then it is easy to see which one is larger: just compare their numerators:

$$\frac{3}{7} < \frac{4}{7} \quad \text{because} \quad 3 < 4.$$

When the denominators of two fractions are different, we have to write them with a common denominator in order to compare them.

Example 1   **Compare the Size of Two Fractions**

Which is larger, $\frac{2}{3}$ or $\frac{5}{8}$?

The least common multiple of 3 and 8 is 24, so we change each fraction into an equivalent fraction with a denominator of 24.

$$\frac{2}{3} = \frac{16}{24} \quad \text{and} \quad \frac{5}{8} = \frac{15}{24}.$$

Since $16 > 15$, we know that $\frac{16}{24} > \frac{15}{24}$, and this means that

$$\frac{2}{3} > \frac{5}{8}.$$

There is an easy way to do the comparison in Example 1 called the **cross product rule**. Write the two fractions side by side and multiply across and up as shown below.

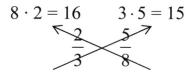

$$8 \cdot 2 = 16 \qquad 3 \cdot 5 = 15$$

Multiplying $8 \cdot 2$ gives the numerator 16 of the fraction $\frac{16}{24}$ that is equivalent to $\frac{2}{3}$. Multiplying $3 \cdot 5$ gives the numerator 15 of the fraction $\frac{15}{24}$ that is equivalent to $\frac{5}{8}$. So we immediately see that $\frac{2}{3} > \frac{5}{8}$ because $16 > 15$.

Here is the general rule:

Cross Product Rule

$$\frac{a}{b} < \frac{c}{d} \quad \text{if and only if} \ ad < bc,$$

$$\frac{a}{b} = \frac{c}{d} \quad \text{if and only if} \ ad = bc, \ \text{and}$$

$$\frac{a}{b} > \frac{c}{d} \quad \text{if and only if} \ ad > bc.$$

Example 2   Compare the Size of Two Fractions

Which is larger, $\frac{2}{3}$ or $\frac{12}{18}$?

$$18 \cdot 2 = 36 \qquad 3 \cdot 12 = 36$$

Since both cross products equal 36, we have $\frac{2}{3} = \frac{12}{18}$.

Example 3   Compare the Size of Two Fractions

Which is larger, $\frac{5}{7}$ or $\frac{3}{4}$?

$$4 \cdot 5 = 20 \qquad 7 \cdot 3 = 21$$

$$\frac{5}{7} \times \frac{3}{4}$$

Since $21 > 20$, we see that $\frac{3}{4}$ is larger than $\frac{5}{7}$.

One important property of the counting numbers is the idea of "What comes next?" But for rational numbers, there is no next larger number. For example, there is no fraction that is next larger than $\frac{2}{3}$. The fraction $\frac{3}{4}$ is slightly larger, but our next example shows how to get an infinite list of numbers each of which is larger than $\frac{2}{3}$ and smaller than $\frac{3}{4}$.

Example 4   Finding Intermediate Fractions

Find two fractions that are between $\frac{2}{3}$ and $\frac{3}{4}$.

Let's write some equivalent fractions for $\frac{2}{3}$ and $\frac{3}{4}$.

$$\frac{2}{3} = \frac{8}{12} = \frac{16}{24} = \frac{24}{36} = \frac{32}{48} = \cdots$$

$$\frac{3}{4} = \frac{9}{12} = \frac{18}{24} = \frac{27}{36} = \frac{36}{48} = \cdots$$

Study Tip:

First find equivalent fractions with a common denominator. Then look at multiples.

If we look at the common denominator of 24 we find one fraction, $\frac{17}{24}$, that is between $\frac{2}{3}$ and $\frac{3}{4}$.

$$\frac{2}{3} = \frac{16}{24} < \frac{17}{24} < \frac{18}{24} = \frac{3}{4},$$

If we look at the common denominator of 36 we find two fractions, $\frac{25}{36}$ and $\frac{26}{36}$, that are between $\frac{2}{3}$ and $\frac{3}{4}$.

$$\frac{2}{3} = \frac{24}{36} < \frac{25}{36} < \frac{26}{36} < \frac{27}{36} = \frac{3}{4}$$

We could continue in the same manner to get an infinite list of fractions between $\frac{2}{3}$ and $\frac{3}{4}$.

EXERCISE 7.5

DEVELOP YOUR SKILL

In Exercises 1 – 6, indicate which fraction is larger.

1. $\frac{4}{9}$ or $\frac{2}{5}$　　**2.** $\frac{2}{7}$ or $\frac{1}{3}$　　**3.** $\frac{2}{5}$ or $\frac{5}{13}$　　**4.** $\frac{3}{7}$ or $\frac{5}{12}$

5. $\frac{6}{11}$ or $\frac{5}{9}$　　**6.** $\frac{4}{11}$ or $\frac{3}{8}$　　**7.** $\frac{4}{5}$ or $\frac{5}{7}$　　**8.** $\frac{11}{7}$ or $\frac{5}{3}$

Find two fractions that are between each pair of fractions.

9. $\frac{1}{5}$ and $\frac{1}{4}$　　**10.** $\frac{1}{2}$ and $\frac{2}{3}$　　**11.** $\frac{2}{3}$ and $\frac{5}{7}$　　**12.** $\frac{3}{4}$ and $\frac{4}{5}$

MAINTAIN YOUR SKILL

Name each form and then compute the basic numeral. You may use fractions in your answers. [6.2]

13. $3 \cdot 6 - 4 \cdot 2$　　　　**14.** $(9 - 2 \cdot 3)^2$　　　　**15.** $16 \div 2^2$

16. $3 + 7 \div 2$ **17.** $\dfrac{6 \cdot 7^0}{3}$ **18.** $5\left(\dfrac{8}{3} - 1\right)$

Evaluate each expression when $x = \dfrac{1}{2}$ and $y = \dfrac{2}{3}$. [6.2, 7.2, 7.4]

19. $5x^2 - y$ **20.** $x + 3y^2$

Solve each equation by using an invariant principle. You may use fractions in your answers. [3.8, 3.9]

21. $32 \cdot 12 = 8 \cdot x$ **22.** $12 \cdot 7 = 36 \cdot x$

23. $\dfrac{13}{22} = \dfrac{x}{66}$ **24.** $\dfrac{35}{9} = \dfrac{7}{x}$

CHAPTER 7 REVIEW

Describe the effect of each operator as the composition of an expansion and a contraction. Then indicate whether the overall effect is to make a nonzero operand larger or smaller. [7.1]

1. $\dfrac{3}{8}(\ )$ **2.** $\dfrac{4}{3}(\ )$

3. $\dfrac{9}{5}(\ )$ **4.** $\dfrac{5}{7}(\ )$

Reduce each fraction to lowest terms. [7.1]

5. $\dfrac{2}{8}$ **6.** $\dfrac{3}{12}$ **7.** $\dfrac{10}{15}$

8. $\dfrac{16}{24}$ **9.** $\dfrac{40}{48}$ **10.** $\dfrac{21}{28}$

Write the reciprocal as a simple fraction or a natural number. [7.2]

11. $\dfrac{7}{8}$ **12.** $\dfrac{10}{3}$ **13.** $\dfrac{1}{9}$

14. 6

15. $5\frac{2}{3}$

16. $4\frac{1}{5}$

Compute the basic numeral. [7.2, 7.3, 7.4]

17. $\left(\frac{3}{8}\right)\left(\frac{2}{3}\right)$

18. $\left(\frac{4}{3}\right)\left(\frac{5}{16}\right)$

19. $\left(\frac{5}{6}\right)\left(\frac{9}{10}\right)$

20. $\left(\frac{5}{6}\right)\left(1\frac{1}{3}\right)$

21. $\frac{5}{8} \div \frac{7}{4}$

22. $\frac{2}{5} \div \frac{4}{3}$

23. $\dfrac{1\frac{3}{8}}{\frac{7}{8}}$

24. $\dfrac{\frac{1}{2}}{1\frac{2}{3}}$

25. $\frac{3}{5} + \frac{7}{10}$

26. $\frac{3}{4} - \frac{1}{6}$

27. $\frac{7}{8} - \frac{1}{3}$

28. $\frac{7}{10} + \frac{2}{15}$

29. $4\frac{2}{3} + 5\frac{1}{2}$

30. $12\frac{3}{4} + 6\frac{1}{3}$

Solve each equation for x and compute the basic numeral. [7.3, 7.4]

31. $x + \frac{3}{8} = \frac{5}{4}$

32. $\frac{5}{4} - x = \frac{1}{3}$

33. $\frac{7}{8}x = \frac{3}{4}$

34. $\frac{3}{8} \div x = \frac{5}{4}$

Indicate which fraction is larger. [7.5]

35. $\frac{3}{5}$ or $\frac{5}{8}$

36. $\frac{5}{11}$ or $\frac{4}{9}$

Find two fractions that are between each pair of fractions. [7.5]

37. $\frac{3}{5}$ and $\frac{2}{3}$

38. $\frac{4}{5}$ and $\frac{5}{6}$

Chapter 8 – Decimals and Signed Numbers

8.1 – DECIMAL FRACTIONS

A **decimal** (or a **decimal fraction**) is a shorthand way of writing a fraction with a denominator that is a power of 10. The number of zeros in the denominator tells us how many digits should be to the right of the decimal point. For example,

$$\frac{27}{10} = 2.7$$ There is one zero in the denominator and one digit to the right of the decimal point.

$$\frac{1538}{100} = 15.38$$ There are two zeros in the denominator and two digits to the right of the decimal point.

$$\frac{5}{1000} = 0.005$$ There are three zeros in the denominator and three digits to the right of the decimal point.

In the last example we added two zeros to the left of the 5 as placeholders so that there could be three digits to the right of the decimal point. Then we added one more zero so that the number would not begin with the decimal point.

The quotient notation for rational numbers is called the **common fraction** notation. A number in this form is often just called a **fraction**. To change from the decimal form into the common fraction form, first rewrite the decimal as a common fraction where the denominator is the appropriate power of ten. Then reduce, if needed

Example 1   Write a Decimal as a Fraction

$1.3 = \dfrac{13}{10}$ There is one digit to the right of the decimal point, so the denominator is 10.

$0.043 = \dfrac{43}{1000}$ There are three digits to the right of the decimal point, so the denominator is 1000.

$0.15 = \dfrac{15}{100} = \dfrac{3}{20}$ There are two digits to the right of the decimal point, so the denominator is 100. Then reduce.

$2.5 = \dfrac{25}{10} = \dfrac{5}{2}$ There is one digit to the right of the decimal point, so the denominator is 10. Then reduce.

$2.50 = \dfrac{250}{100} = \dfrac{25}{10} = \dfrac{5}{2}$ There are two digits to the right of the decimal point, so the denominator is 100. Then reduce.

Study Tip:

Note that
2.5 = 2.50
Adding one or more zeros at the right end of a decimal does not change its value.

Example 2   Divide a Decimal by a Natural Number

Compute the basic numeral: $2.1 \div 3$

$2.1 \div 3 = \dfrac{2.1}{3} = \dfrac{\frac{21}{10}}{3}$ Write the decimal fraction as a common fraction.

$= \dfrac{\frac{21}{3}}{10} = \dfrac{7}{10}$ Interchange the dividing by 10 and the dividing by 3. Then compute.

$= 0.7$ Change the common fraction back into a decimal fraction.

If we write this out as long division we have

$$
\begin{array}{r}
0.7 \\
3\overline{)2.1} \\
\underline{2\,1} \\
0
\end{array}
$$

Note that the decimal point in the dividend 2.1 goes straight up to become the decimal point in the quotient 0.7. This is because the denominator of the final fraction is still 10.

To write a common fraction as a decimal fraction, divide the denominator into the numerator. In simple cases, you can use long division. For larger numbers, use a calculator.

Study Tip:

We usually write
    0.375
instead of
    .375
because the decimal point is easy to miss. Having a 0 to the left of the decimal point calls attention to it.

Example 3   Write a Fraction as a Decimal

Write $\dfrac{3}{8}$ as a decimal fraction.

We begin by writing 3 as 3.000. Then use long division to divide this by 8:

$$
\begin{array}{r}
0.375 \\
8\overline{)3.000} \\
\underline{2\,4} \\
60 \\
\underline{56} \\
40 \\
\underline{40} \\
0
\end{array}
$$

Division ends when the remainder is 0.

The fraction $\dfrac{3}{8}$ is equivalent to the decimal 0.375.

If the denominator of a fraction can easily be changed into a power of 10, then the fraction can be converted into a decimal without long division.

Study Tip:

For a natural number like 46, we can think of the decimal point as being at the right end. So dividing by 100 has the effect of moving the decimal point 2 places to the left.

Example 4   **Write a Fraction as a Decimal**

Write $\dfrac{23}{50}$ *as a decimal.*

Multiply by $\dfrac{2}{2}$ to change the denominator into 100.

$$\frac{23}{50} = \frac{23 \times 2}{50 \times 2} = \frac{46}{100} = 0.46$$

EXERCISE 8.1

DEVELOP YOUR SKILL

Change each decimal into an equivalent common fraction in lowest terms.

1. 0.29   **2.** 4.3   **3.** 0.046

4. 0.35   **5.** 2.6   **6.** 1.45

Change each common fraction into an equivalent decimal.

7. $\dfrac{37}{100}$ **8.** $\dfrac{21}{10}$ **9.** $\dfrac{3}{5}$

10. $\dfrac{7}{5}$ **11.** $\dfrac{9}{25}$ **12.** $\dfrac{11}{50}$

Compute the basic numeral. Write the answer as a decimal.

13. $5.16 \div 3$ **14.** $17.4 \div 2$

15. $13.5 \div 5$ **16.** $2.36 \div 4$

MAINTAIN YOUR SKILL

Solve each equation for x. [7.3, 7.4]

17. $\dfrac{7}{8} - x = \dfrac{1}{6}$

18. $\dfrac{7}{8} \div x = \dfrac{1}{6}$

19. $\dfrac{3}{4} + x = \dfrac{13}{8}$

20. $\left(\dfrac{3}{4}\right)(x) = \dfrac{13}{8}$

21. $x - \dfrac{4}{5} = \dfrac{7}{10}$

22. $x \div \dfrac{4}{5} = \dfrac{7}{10}$

Indicate which fraction is larger. [7.5]

23. $\dfrac{4}{7}$ or $\dfrac{5}{9}$

24. $\dfrac{5}{8}$ or $\dfrac{7}{11}$

8.2 – COMPUTING WITH DECIMAL FRACTIONS

OBJECTIVE

1. Compute with decimal fractions.

Computing with decimal fractions is the same as computing with natural numbers, except that we have to keep track of where the decimal point goes. Let's look at some simple examples to see how to do this. We begin by using the definition of a decimal and then show how to do it easily by writing the problem vertically.

Example 1   Add Two Decimal Fractions

Compute the sum: $14.5 + 0.23$

We begin by changing each decimal fraction into a common fraction, leaving the denominators as powers of ten.

$$14.5 = \frac{145}{10} \quad \text{and} \quad 0.23 = \frac{23}{100}$$

So, $14.5 + 0.23 = \dfrac{145}{10} + \dfrac{23}{100}$ Change the decimals into fractions.

$$= \frac{1450}{100} + \frac{23}{100}$$ Get a common denominator.

$$= \frac{1473}{100} = 14.73$$ Add and change back into a decimal.

To write this out vertically, we align the decimal points in the addends and the sum:

$$
\begin{array}{r}
14.50 \\
+\ \ 0.23 \\
\hline
14.73
\end{array}
$$

Example 2   Add and Subtract Decimal Fractions

Compute the sum and the difference of 26.48 *and* 2.7.

We write both computations vertically and align the decimal points.

$$
\begin{array}{r}
26.48 \\
+\ \ 2.70 \\
\hline
29.18
\end{array}
\qquad
\begin{array}{r}
2\overset{5}{\cancel{6}}.\overset{1}{4}8 \\
-\ \ 2.70 \\
\hline
23.78
\end{array}
$$

Study Tip:

We wrote 2.7 as 2.70 so both numbers would have two digits to the right of the decimal point.

When we multiply two decimal fractions, we don't have to align the decimal points like we do with adding and subtracting. But we do have to keep track of how many digits are to the right of the decimal point in each factor.

Example 3   Multiply Two Decimal Fractions

Compute the product: 12.3×0.24

We begin by changing each decimal fraction into a common fraction. To multiply the common fractions, we multiply the numerators and multiply the denominators.

$$12.3 \times 0.24 = \frac{123}{10} \times \frac{24}{100} = \frac{123 \times 24}{1000}$$

We see that there is one digit to the right of the decimal point in the first factor and two digits to the right of the decimal point in the second factor. The denominator of the product is 1000. This means there will be three digits to the right of the decimal point in our answer. This 3 comes from adding the 1 and the 2.

So to compute the product of the decimals 12.3 and 0.24, we compute the product of the natural numbers 123 and 24 and then insert the decimal point in the right place.

$$
\begin{array}{r}
12.3 \\
\times\ 0.24 \\
\hline
492 \\
246 \\
\hline
2.952
\end{array}
$$

12.3 ⟵ 1 digit to the right of the decimal point.

× 0.24 ⟵ 2 digits to the right of the decimal point.

2.952 ⟵ 3 digits to the right of the decimal point.

Example 4   **Multiply Two Decimal Fractions**

Compute the product: 31.8×0.4

$$
\begin{array}{r}
31.8 \\
\times \ \ 0.4 \\
\hline
12.72
\end{array}
$$

31.8   ⟵ 1 digit to the right of the decimal point.

× 0.4   ⟵ 1 digit to the right of the decimal point.

12.72   ⟵ 2 digits to the right of the decimal point.

Example 5   **Divide Two Decimal Fractions**

Compute the quotient: $17.91 \div 0.3$

We begin by using the invariant principle for quotients to multiply the dividend and the divisor both by 10. This changes the divisor into a natural number.

$$\frac{17.91}{0.3} = \frac{17.91 \times 10}{0.3 \times 10} = \frac{179.1}{3}$$

Now the problem is like Example 2 in Lesson 8.1. Using long division, it looks like this:

Study Tip:

Multiplying the dividend and the divisor by 10 moves their decimal points one place to the right.

$$0.3\overline{)17.91} \longrightarrow$$

$$
\begin{array}{r}
59.7 \\
3\overline{)179.1} \\
15 \\
\hline
29 \\
27 \\
\hline
2\,1 \\
2\,1 \\
\hline
0
\end{array}
$$

Example 6   **Divide Two Decimal Fractions**

Compute the quotient: $18.3 \div 0.02$

In order to change the divisor into a natural number, we multiply it by 100. Of course we must also multiply the dividend by the same amount.

$$\frac{18.3}{0.02} = \frac{18.3 \times 100}{0.02 \times 100} = \frac{1830}{2}$$

Using long division, it looks like this:

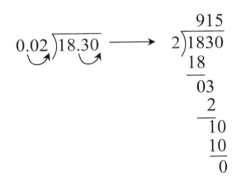

Study Tip:

Multiplying the dividend and the divisor by 100 moves their decimal points two places to the right.

EXERCISE 8.2

DEVELOP YOUR SKILL

Compute the basic numeral. Write the answer as a decimal.

1. $32.6 + 3.17$
2. $0.058 + 0.39$
3. $16.73 - 4.5$
4. $23.8 - 4.52$
5. 21.8×0.4
6. 0.13×0.5
7. 31.4×0.3
8. 26.8×0.07
9. $28.8 \div 0.3$
10. $39.2 \div 0.7$
11. $2.34 \div 0.6$
12. $37.5 \div 0.05$

MAINTAIN YOUR SKILL

13. The pizza was cut into eight equal slices. After Jane, Carla, and Laura each took a slice, what fraction of the pizza was left? [7.4]

In Exercises 14 – 19, each equation can be viewed as a sum and two addends or as a product and two factors. Identify the role of x in each equation. For example, in the equation $x + 3 = k$, x is an addend. In the equation $4x = c$, x is a factor. [5.2, 5.4]

14. $x - 3 = y$ **15.** $b = 40 \div x$

16. $32 = x \div n$ **17.** $n = 18 - x$

18. $15 + x = a$ **19.** $k = 5x$

Solve each equation for x. These are the same equations used in Exercises 14 – 19. [5.2, 5.4]

20. $x - 3 = y$ **21.** $b = 40 \div x$

22. $32 = x \div n$ **23.** $n = 18 - x$

24. $15 + x = a$ **25.** $k = 5x$

8.3 – ROUNDING AND ESTIMATING

OBJECTIVE

1. Estimate the sum and difference of fractions and decimals.

The number 23 is 3 more than 20, or 2 tens. It is 7 less than 30, or 3 tens. Suppose that 23 is a count or measure that we wish to record only to the nearest multiple of 10. Since 23 is between 20 and 30, but is nearest to 20, it can be **rounded** to 20. The numbers 21, 22, 23, and 24 may all be "rounded down" to 20.

The numbers 26, 27, 28, and 29 are all closer to 30 than to 20 (or any other multiple of 10). So they could be "rounded up" to 30. The number 25 is halfway between 20 and 30. Numbers ending in 5 are usually rounded up to the larger round number, so 25 would also round up to 30.

Study Tip:

When a natural number is rounded, the "round number" ends in one or more zeros. That's why a **round number** is called "round."

Example 1   Rounding a Number

Round 23,754 *as indicated.*

(a) To the nearest 10   (b) To the nearest 100
(c) To the nearest 1,000.   (d) To the nearest 10,000.

(a) 54 rounds to 50, so 23,754 rounds to 23,750.

(b) The hundreds' digit is a 7 and the digit to the right of 7 is 5, so 754 rounds to 800. This means that 23,754 rounds to 23,800.

(c) The thousands' digit is 3 and the digit to the right of 3 is 7. This means that 23,754 rounds up to 24,000.

(d) The ten-thousands' digit is 2 and the digit to the right of 2 is 3. So 23,754 rounds down to 20,000.

Decimal fractions and mixed numbers can also be rounded to the nearest natural number. If the fractional part is less than one-half, then the number is rounded down. If the fractional part is one-half or more, then the number is rounded up.

Example 2   Rounding Decimals

Round each decimal to the nearest natural number.

(a) 4.3   (b) 7.5   (c) 6.19

(a) 3 tenths is less than one-half, so 4.3 rounds down to 4.

(b) 5 tenths is equal to one-half, so 7.5 rounds up to 8.

(c) 0.19 is less than 0.50, so 6.19 rounds down to 6.

Example 3   Rounding Mixed Numbers

Round each mixed number to the nearest natural number.

(a) $6\dfrac{2}{3}$ (b) $3\dfrac{4}{9}$ (c) $5\dfrac{1}{2}$

(a) $\dfrac{2}{3} > \dfrac{1}{2}$, so $6\dfrac{2}{3}$ rounds up to 7.

(b) $\dfrac{4}{9} < \dfrac{1}{2}$, so $3\dfrac{4}{9}$ rounds down to 3.

(c) $5\dfrac{1}{2}$ rounds up to 6.

When an exact answer is not needed in a problem, we can estimate the answer by using round numbers instead of exact numbers to do the arithmetic. In estimating, we replace an exact number by a round number that is close to it in order to make the computations easier. Estimating is also a good way to find if an exact answer is reasonable.

Study Tip:

The wavy equal sign "≈" means "approximately equal."

Example 4   Estimate a Sum

Estimate the sum of 413 and 296.

To estimate, we first change the exact numbers to round numbers. 413 is a little more than 400 and 296 is about 300. So,

$$413 + 296 \approx 400 + 300 = 700.$$

We estimate the sum is about 700.

Example 5   **Estimate a Sum or Difference**

Estimate each sum or difference.

(a) $395 + 207$ (b) $7.94 - 2.18$

(c) $9\frac{3}{10} + 5\frac{7}{8}$ (d) $11.7 - 2\frac{4}{5}$

(a) 395 rounds up to 400 and 207 rounds down to 200.

$$395 + 204 \approx 400 + 200 = 600.$$

(b) 7.94 rounds up to 8 and 2.18 rounds down to 2.

$$7.94 - 2.18 \approx 8 - 2 = 6$$

(c) $9\frac{3}{10}$ rounds down to 9 and $5\frac{7}{8}$ rounds up to 6.

$$9\frac{3}{10} + 5\frac{7}{8} \approx 9 + 6 = 15$$

(d) 11.7 rounds up to 12 and $2\frac{4}{5}$ rounds up to 3.

$$11.7 - 2\frac{4}{5} \approx 12 - 3 = 9$$

EXERCISE 8.3

DEVELOP YOUR SKILL

In Exercises 1 – 4, round 48,537 as indicated.

1. To the nearest 10   **2.** To the nearest 100

3. To the nearest 1,000   **4.** To the nearest 10,000

In Exercises 5 – 8, round 126,081 as indicated.

5. To the nearest 10   **6.** To the nearest 100

7. To the nearest 1,000   **8.** To the nearest 10,000

In Exercises 9 – 12, round each number to the nearest natural number.

9. 5.6

10. 3.4

11. $9\frac{1}{2}$

12. $4\frac{3}{7}$

Estimate each sum or difference.

13. $507 + 288$

14. $8.9 - 3.27$

15. $7\frac{4}{5} + 3\frac{1}{6}$

16. $8\frac{3}{8} - 2.37$

MAINTAIN YOUR SKILL

17. Mickey had $20 before he went shopping. After spending money on groceries, he had $2.35 left. Let x be the amount that Mickey spent on groceries. Write a subtraction equation with x as the variable to model this problem. Then solve the equation to find how much Mickey spent on groceries. [5.7, 8.2]

Compute the basic form for each composite operator. [2.2, 2.4]

18. $-5 + 12$

19. $+3 - 12$

20. $-8 - 4$

21. $+15 - 2$

22. $-13 + 8$

23. $+3 + 7$

Evaluate each expression when $x = \dfrac{3}{2}$ and $y = 4$. [6.2, 7.2, 7.4]

24. $2x^2 - y$

25. $x^2 y - xy^0$

8.4 – SIGNED NUMBERS AND THEIR SUMS

OBJECTIVES

1. Introduce the signed numbers.

2. Add signed numbers by using increases and decreases.

The extension of the number system from the counting numbers to the quotients, as in Lesson 7.1, still leaves the difference $x - y$ without meaning when $x < y$. We now extend the number system again to overcome this limitation. A method for doing this is suggested by the vector model that we developed in Lessons 2.2 and 2.4 for increases and decreases.

Example 1   **Vector Models for Increases and Decreases**

For each pair of operators, construct a vector model that shows the two given operators and their composite operator.

  (a) $+ 5, - 2$ (b) $+ 3, - 7$

(a)   The increase $+ 5$ and the decrease $- 2$ work against each other to give a composite change of $+ (5 - 2) = + 3$.

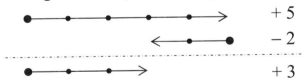

(b)   This time the increase $+ 3$ and the decrease $- 7$ give a composite change of $- 4$ (a decrease of 4).

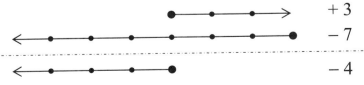

If we write the vector model from Example 1 part (a) underneath a linear scale and have the first operator, + 5, start at zero, then we find that the vector for the composite change, + 3, goes from zero to the number 3.

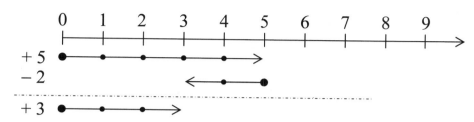

If, however, we try to follow this same pattern in Example 1 part (b), we have a problem since the composite change is a decrease. We can eliminate this difficulty if we extend our scale to the left of zero and give each point a name that corresponds to the decrease operator that begins at zero and ends at the point:

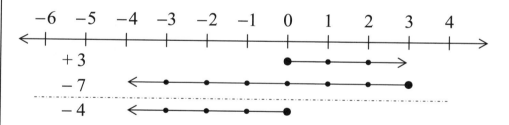

Study Tip:

We may use "+ n" and "+n" interchangeably. Sometimes the "+ n" notation (with more space between the plus sign and the numeral) is used to emphasize the operator point of view and "+n" is used to emphasize its role as a signed number, but we may interpret them in either way that is helpful. Similar comments apply to "− n" and "−n."

Using this extended number line, we have a system in which subtraction is always defined. The numbers to the right of zero are now called **positive numbers**. Those on the left of zero are called **negative numbers**. Zero is neither positive nor negative. Addition operators such as + n can be used as numerals for the positive numbers, but it is easier to continue using the familiar basic numeral. So n and + n refer to the same positive number. From now on, + n and − n will be used both as operators (addition and subtraction) and as numerals for **signed numbers**.

The extended rational number system that includes positive and negative rational numbers and zero is usually denoted by $\mathbb{Q}$ (for quotients) and we have new names for the natural numbers and the counting numbers.

- The natural numbers are now called the **positive integers**.

- The counting numbers are called the **nonnegative integers**.

- The opposites of the positive integers are called the **negative integers**.

- The **integers** consist of the positive integers, the negative integers, and zero.

- The **rational numbers** consist of the positive rational numbers, the negative rational numbers, and zero.

Addition with signed numbers follows the same pattern as the composition of addition and subtraction operators. The positive numbers correspond to increases and the negative numbers correspond to decreases.

To add signed numbers,

Adding Signed Numbers	**Think:** increases and decreases (using $+\,n$ for n) **See:** vectors, as in Chapter 2 **Write:** numbers, guided by the rules for composition of addition and subtraction operators.

Example 2   Compute the Sum of Signed Numbers

Compute the sum of negative 6 *and positive* 4:

$$(-6) + (+4)$$

We want to think of this as the composition of a decrease of 6 and an increase of 4. Since 6 is larger than 4, the combined change is a decrease.

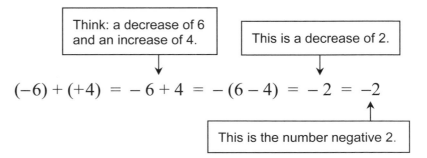

> Think: a decrease of 6 and an increase of 4.

> This is a decrease of 2.

$$(-6) + (+4) \ = \ -6 + 4 \ = \ -(6-4) \ = \ -2 \ = \ -2$$

> This is the number negative 2.

We can see this in our mind by visualizing the vectors:

$$
\begin{array}{l}
-6 \\
+4 \\
\hline
-2
\end{array}
$$

Example 3   Compute the Sum of Signed Numbers

Compute the sum of positive 6.7 *and negative* 3.5:

$$(6.7) + (-3.5)$$

We think of this as an increase of 6.7 followed by a decrease of 3.5. Since the increase is larger, the combined change is an increase.

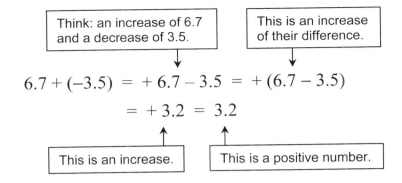

Think: an increase of 6.7 and a decrease of 3.5.

This is an increase of their difference.

$$6.7 + (-3.5) \ = \ +6.7 - 3.5 \ = \ +(6.7 - 3.5)$$
$$= \ +3.2 \ = \ 3.2$$

This is an increase.

This is a positive number.

Example 4   Compute the Sum of Signed Numbers

Compute the sum of the numbers $-\frac{1}{2}$ *and* $-\frac{2}{3}$.

Think of these negative numbers as decreases. The composition of two decreases is a larger decrease.

This is two decreases.

This is a decrease of their sum.

$$-\frac{1}{2} + \left(-\frac{2}{3}\right) = -\frac{1}{2} - \frac{2}{3} = -\left(\frac{1}{2} + \frac{2}{3}\right) \quad \leftarrow \text{The LCD is 6.}$$
$$= -\left(\frac{3}{6} + \frac{4}{6}\right) = -\frac{7}{6} = -\frac{7}{6}$$

This is a decrease.

This is a negative number.

It is not so easy to draw exact vector models when the magnitudes are not integers, but even with fractions, the idea is the same. The two decreases will be represented by vectors pointing to the left, and their composition will be a longer vector pointing to the left. Without the scale markings we can visualize it like this:

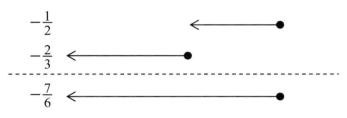

EXERCISE 8.4

DEVELOP YOUR SKILL

Compute the sum of each pair of numbers.

1. -3 and 8

2. 4 and -6

3. -5 and -7

4. $-\frac{1}{3}$ and $\frac{3}{7}$

5. $\frac{2}{3}$ and $-1\frac{1}{4}$

6. $-\frac{1}{2}$ and $-\frac{2}{5}$

7. -2.1 and -3.5

8. 5.4 and -2.7

9. -5.7 and 3.4

10. $1\frac{1}{2}$ and $-\frac{2}{3}$

MAINTAIN YOUR SKILL

Change each product or quotient into an equivalent power. Do not compute, except for exponents. [6.3]

11. $5^4 \times \dfrac{5^3}{5^0}$

12. $\dfrac{5^2 \cdot 5 \cdot 5^2}{5^3}$

Write each number in prime factored form. [6.4]

13. 42

14. 56

Find the GCF and the LCM of each set of numbers. Write your answers in prime factored form. [6.5]

15. $\begin{cases} 3 \cdot 5^2 \cdot 7 \\ 3^2 \cdot 7 \end{cases}$

16. $\begin{cases} 2^3 \cdot 3 \cdot 5^2 \\ 2^2 \cdot 3^2 \cdot 7 \end{cases}$

Solve each equation for x. [5.7, 7.3, 7.4, 8.2]

17. $x - 2.5 = 4.8$

18. $3.6 = 15.01 - x$

19. $\dfrac{13.5}{x} = 3$

20. $5x = 12.35$

21. $\left(1\dfrac{1}{2}\right)(x) = 2\dfrac{1}{2}$

22. $\dfrac{2\frac{1}{3}}{x} = 4\dfrac{1}{3}$

23. $3\dfrac{1}{2} - x = \dfrac{3}{4}$

24. $x + \dfrac{2}{3} = 1\dfrac{3}{8}$

8.5 – INEQUALITIES AND THEIR GRAPHS

OBJECTIVES

1. Graph an inequality.

2. Solve a simple inequality.

In our original definition of inequality in Lesson 1.3, to show that $x < y$, we needed a natural number n that could be added on to x to make the sum equal to y. Recall,

$x < y$ if there is a natural number n such that $x + n = y$.

We can extend this definition to signed numbers by requiring that n be any number greater than 0.

Inequalities with Signed Numbers

Let x and y be signed numbers. We say that $x < y$ if there is a positive number n such that $x + n = y$. Similarly, $x > y$ if there is a positive number n such that $x = y + n$.

When an inequality involves a variable, there are usually many possible solutions. For example, the inequality $x > 3$ is satisfied by 5, $\frac{22}{3}$, 138, and many other numbers. One way to display the solutions of an inequality is to graph them on a number line. If the endpoint of a line is to be included in the graph, it is indicated by a closed (solid) dot. If the endpoint is not included, an open (hollow) dot is used.

Example 1   Graph an Inequality

Graph each inequality on a number line.
   (a) $x > 3$ (b) $x \le 5$

(a) The numbers that are greater than 3 are to the right of 3 on the number line. Since 3 is not included, an open dot is used at the left endpoint. The arrow pointing to the right indicates that any number to the right of 3 is a solution.

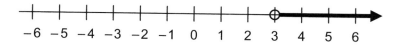

(b) The numbers that are less than or equal to 5 are to the left of 5 on the number line. Since 5 is included, a closed dot is used at the right endpoint.

Example 2   Check an Inequality

Which of the following numbers are solutions to the inequality $x < 2$?

   (a) $\dfrac{13}{3}$ (b) 2.53   (c) $\dfrac{10}{7}$

         (d) –3   (e) – 0.4

(a) $\dfrac{13}{3}$ is equal to $4\dfrac{1}{3}$ and this is a little more than 4. So $\dfrac{13}{3}$ is larger than 2 and it is not a solution.

(b) 2.53 is more than 2 and it is not a solution.

(c) $\dfrac{10}{7}$ is equal to $1\dfrac{3}{7}$, and this is between 1 and 2. So $\dfrac{10}{7}$ is a solution.

(d) 3 is larger than 2, but –3 is less than 2. We see from our definition of inequality that $-3 < 2$ because $-3 + 5 = 2$. So –3 is a solution.

(e) Every negative number is less than 0. This means every negative number is less than any positive number. So – 0.4 is a solution.

   Let's graph the inequality $x < 2$ and show the numbers from Example 2 on the same graph. We see that $-3, - 0.4$, and $1\dfrac{3}{7}$ are to the left of 2 and 2.53 and $4\dfrac{1}{3}$ are to the right of 2.

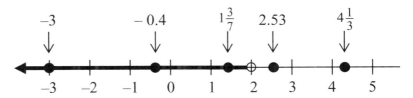

An inequality like $1 < 5$ says that 1 is to the left of 5 on the number line. If the same number, say 2, is added to 1 and to 5, the resulting values maintain the same relationship: $3 < 7$. Both numbers have just been translated to the right by 2 units.

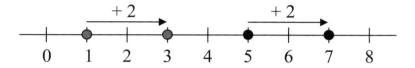

This illustrates that an inequality remains valid when both sides are increased (or decreased) by the same amount.

Example 3   Solve an Inequality

Solve $x + 6 \geq 10$ *and graph the solution..*

$$x + 6 \geq 10$$ Subtract 6 from both sides. Or, remove the increase of 6 from the left side and join a decrease of 6 to the right side.

$$x \geq 10 - 6$$

$$x \geq 4$$ Compute.

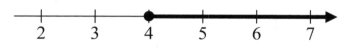

Caution

In Chapter 5 we introduced the 3-number method of solving equations. Do <u>not</u> use this method when solving inequalities. It doesn't always give the correct direction for the inequality.

EXERCISE 8.5

DEVELOP YOUR SKILL

Indicate which of the given numbers are solutions to the inequality.

1. $x < 4;$ $\dfrac{11}{2}, \dfrac{11}{3}, -2, 3.9$

2. $x > -3;$ $-2, -5, -\dfrac{7}{2}, \dfrac{1}{3}$

Solve each inequality and graph the solution.

3. $x + 6 < 8$

4. $y - 3 \leq 5$

5. $n - 2 \geq 5$

6. $m + 4 > 9$

7. $x - 6 < -2$

8. $y - 2 \geq -5$

9. $n + 3 > 11$

10. $m + 6 \leq 8$

MAINTAIN YOUR SKILL

Analyze each expression by stating a basic operand and then listing a sequence of basic operators that can be joined one at a time to produce the given expression. If more than one operand and sequence of operators can be used, give the alternate possibilities. [4.5]

11. $\dfrac{17 - 2}{5}$

12. $a + \dfrac{b}{3}$

13. $\dfrac{12 - x}{y} + 3$

14. $5\left(\dfrac{16}{2} + c\right)$

Name each form and compute the basic numeral. [6.2]

15. $18 - 4 \cdot 3$

16. $6 \cdot 4^0$

17. $6\left(4 - \dfrac{1}{3}\right)$

18. $4 \cdot 6 - 2$

19. $(2^3 - 2)^2$

20. $\dfrac{5 \cdot 2}{5 \cdot 2^2}$

In Exercises 21 – 24, round 253,408 as indicated. [8.3]

21. To the nearest 10

22. To the nearest 100

23. To the nearest 1,000

24. To the nearest 10,000

CHAPTER 8 REVIEW

Change each decimal into an equivalent common fraction in lowest terms and change each common fraction into an equivalent decimal. [8.1]

1. 0.37   **2.** 5.2   **3.** $\dfrac{57}{10}$ **4.** $\dfrac{12}{25}$

Compute the basic numeral. Write the answer as a decimal. [8.2]

5. $17.3 + 25.8$ **6.** $1.45 + 3.7$

7. $35.7 - 26.52$ **8.** $6.02 - 3.4$

9. 0.43×1.6 **10.** 2.5×0.03

11. $9.36 \div 0.02$ **12.** $0.032 \div 0.4$

In Exercises 13 – 16, round 48,275 as indicated. [8.3]

13. To the nearest 10   **14.** To the nearest 100

15. To the nearest 1,000   **16.** To the nearest 10,000

Estimate each sum or difference. [8.3]

17. $693 + 218$ **18.** $12\dfrac{7}{8} - 3\dfrac{2}{7}$

19. $17.2 - 5.834$ **20.** $8\dfrac{5}{6} + 4.09$

Compute the sum of each pair of numbers. [8.4]

21. -5 and 7 **22.** -3 and -6

23. $\dfrac{1}{2}$ and $-\dfrac{2}{3}$ **24.** $-2\dfrac{2}{3}$ and $-1\dfrac{3}{4}$

25. 8.5 and -2.13 **26.** -3.41 and -12.5

Solve each equation and graph the solution set. [8.5]

27. $x - 5 < 3$ **28.** $x + 4 \geq 5$

29. $y + 2 \leq -1$ **30.** $y - 7 > -4$

Glossary of Terms

Addend: one of the numbers being added to form a sum. (page 137)	In the sum $3 + 5$, the 3 and the 5 are addends.
Additive Identity: the additive identity is 0 because adding zero makes no change. (page 62)	The additive identity operator is $+ 0$.
Additive Inverse: the opposite of an increase or decrease operator. When an increase or decrease is followed by its additive inverse, the combined effect is no change. (page 62)	The additive inverse of $+ 5$ (an increase of 5) is $- 5$ (a decrease of 5). $$+ 5 - 5 = + 0 \text{ (no change)}$$ The additive inverse of $- 7$ (a decrease of 7) is $+ 7$ (an increase of 7). $$- 7 + 5 = + 0 \text{ (no change)}$$
Associative Principle: the sum (or the product) of three numbers is always the same regardless of their grouping. (pages 24 & 88)	Associative Principle for Addition: $$x + (y + z) = (x + y) + z$$ Associative Principle for Multiplication: $$x \cdot (y \cdot z) = (x \cdot y) \cdot z$$ Note that the order of the addends (and the order of the factors) stays the same.
Basic Numeral: the simplest way of writing a number. For a fraction, it will be reduced to lowest terms. (pages 3 & 202)	The basic numeral for $5 + 3$ is 8. Replacing $5 + 3$ by 8 is called a computation. The basic numeral for $\frac{2}{4}$ is $\frac{1}{2}$.
Basic Operator: the simplest way of writing an operator. (page 41)	The basic operator for $+ 5 + 3$ (an increase of 5 followed by an increase of 3) is $+ 8$ (an increase of 8), since the combined effect of adding 5 and adding 3 is to add 8.

Commutative Principle: the addends in a sum and the factors in a product may be interchanged.
 	
   The commutative principle can also be extended to operators. Increases commute with decreases and expansions commute with contractions.
(pages 22, 86 & 120)
	Commutative Principle for Addition:
	$$x + y = y + x$$
	Commutative Principle for Multiplication:
	$$x \cdot y = y \cdot x$$
	   A decrease of 7 followed by an increase of 3 is the same as an increase of 3 followed by a decrease of 7:
	$$-7 + 3 = +3 - 7$$
	   An expansion by 6 followed by a contraction by 2 is the same as a contraction by 2 followed by an expansion by 6:
	$$\frac{6(\ )}{2} = 6\left(\frac{(\ )}{2}\right)$$

Composite Change: the overall change (or combined change) when two or more operators are applied to an operand. (page 41)	When an increase of 7 is followed by a decrease of 3, the composite change is an increase of 4: $$+7 - 3 = +4$$
Composite Number: a natural number greater than 1 that has a factor other than 1 and itself. (page 187)	The number 4 is composite since 2 is factor of 4. The number 7 is not composite since its only factors are 1 and 7. (7 is called *prime*.)
Composition of Operators: the overall change when two or more operators are combined. (page 41)	The composition of a decrease of 8 and an increase of 3 is a decrease of 5: $$-8 + 3 = -5$$ The composition of a contraction by 8 and an expansion by 2 is a contraction by 4: $$2\left(\frac{(\ )}{8}\right) = \frac{(\ )}{4}$$
Computation: a change of form where the basic numeral replaces any other numeral for a number. (page 3)	Replacing $5 + 3$ by 8 is a computation.

Conditional Equation: an equation that contains a variable (such as x). Whether the equation is true or not depends on the value chosen for the variable. (page 135)	$x + 3 = 10$ is a conditional equation. It is true when $x = 7$, but not true for other values of x.
Contraction Operator: an operator that represents division. (page 74)	Dividing by 2 (or a contraction by 2) is represented by the operator $\dfrac{(\ )}{2}$.
Counting Numbers: the natural numbers and zero. (page 1)	The counting numbers are $$0, 1, 2, 3, 4, 5, \ldots$$
Cross Product Rule: a way to compare the size of two fractions. If the cross products are equal, then the fractions are equal. (page 224)	$8 \cdot 2 = 16 \qquad 3 \cdot 5 = 15$ $\dfrac{2}{3} \quad\times\quad \dfrac{5}{8}$ Since 16 is larger than 15, $\dfrac{2}{3} > \dfrac{5}{8}$.
Decimal (or Decimal Fraction): a shorthand way of writing a fraction with a denominator that is a power of 10. (page 229)	$\dfrac{27}{10} = 2.7, \qquad \dfrac{5}{1000} = 0.005$ The number of zeros in the denominator tells us how many digits are to the right of the decimal point.
Decrease Operator: an operator that represents subtraction. (page 38)	Subtracting 5 (or a decrease of 5) is represented by the operator -5. The operator "-5" is read "a decrease of five" or "minus five." It is <u>not</u> read as "negative five."
Denominator: the part of a fraction that is below the line. It divides into the numerator (on the top). (page 200)	In the fraction $\dfrac{3}{5}$, the denominator is 5.
Difference: the name of the form when the last operation is subtraction. (page 15)	The expression $24 - 3 \cdot 5$ is a difference because you multiply first and then subtract.

Distributive Principle: relates adjacent levels on the order of operations chart. Powers and roots distribute over products and quotients. Products and quotients distribute over sums and differences. (page 92)	Powers distribute over products and quotients: $$(3 \cdot 5)^2 = 3^2 \cdot 5^2 \quad \text{and} \quad \left(\frac{6}{3}\right)^2 = \frac{6^2}{3^2}$$ Products and quotients distribute over sums and differences: $$(2)(3 + 5) = (2)(3) + (2)(5)$$ $$5(x - y) = 5x - 5y$$
Dividend: the number that is being divided in a quotient. It is written on top. (page 77)	The dividend in the quotient $\frac{8}{2}$ is 8.
Divides: y divides x (or y divides *into* x) if there is a unique counting number n such that $y \cdot n = x$. (page 76)	3 divides 12 because $3 \cdot 4 = 12$. If y divides x, then y is a factor of x and x is a multiple of y.
Divisor: the number written below the line in a quotient. The divisors of a number are the same as the factors of the number. (page 77)	The divisor in the quotient $\frac{8}{2}$ is 2. We say that 2 is a divisor of 8 because $2 \cdot 4 = 8$.
Equivalent Equations: equations which have the same truth value as each other (either both true or both false) for any choice of the variables. (page 136)	The equations $2x - 3 = 9$ and $2x = 12$ are equivalent equations. Both equations are true when $x = 6$ and false for any other value of x.
Equivalent Operators: operators that give the same transform when joined to an operand. Equivalent operators produce the same change. (page 41)	The operators $-5 - 3$ and $-(5 + 3)$ are equivalent because they both represent a decrease of 8.
Evaluate: substitute given numerical values for a variable or variables and compute the basic numeral. (page 181)	To evaluate $x^2 + xy$ when $x = 5$ and $y = 7$, we substitute 5 for x and 7 for y and compute: $$x^2 + xy = 5^2 + (5)(7) = 25 + 35 = 60$$

Expansion Operator: an operator that represents multiplication. (page 69)	Multiplying by 2 (or an expansion by 2) is represented by the operator 2( ) or ( )2,
Explicit: an equation is explicit for a variable (or number) if the variable (or number) appears only once in the equation, and it is one of the members by itself. (page 136)	The equation $x = 2y + 3$ is explicit for the variable x. The equation $\frac{6+4}{2} = 5$ is explicit for the number 5.
Exponent: a positive integer exponent counts the number of times 1 is multiplied by the base. (pages 173)	In the expression 5^3, 3 is the exponent and 5 is the base. $5^3 = 1 \cdot 5 \cdot 5 \cdot 5 = 125$
Exponentiation: the operation of raising to a power. (page 173)	The expression $(2+5)^2$ is called a power because the last operation is exponentiation (raising to the second power).
Factor: a number which when multiplied by another number gives a product. (page 68)	3 and 4 are a factors of 12 because $3 \cdot 4 = 12$. 3 and 4 are factors of the product $(3)(4)$. 2 and $(x-y)$ are factors of the product $2(x-y)$. The factors of a number are the same as the divisors of a number.
Form of an Expression: the name given to an expression. It comes from the last operation performed, when following the order of operations. (page 85)	Last Operation / Form: addition / sum; subtraction / difference; multiplication / product; division / quotient; exponentiation / power. $2 + 3 \cdot 5$ is a **sum** because the *last* operation is addition.
Fraction: a number in the form $\frac{m}{n}$, where m and n are counting numbers and $n \neq 0$. A fraction is also called a *rational number*. (page 198)	The fraction $\frac{m}{n}$ is the solution of the equation $(n)(x) = m$. For example, the fraction $\frac{2}{3}$ is the number which when multiplied by 3 is equal to 2.

Greatest Common Factor (GCF): the largest of the common factors of two or more numbers. (page 191)	Factors of 24: 1, 2, 3, 4, **6**, 8, 12, 24 Factors of 30: 1, 2, 3, 5, **6**, 10, 15, 30 The GCF of 24 and 30 is 6.
Identity Operator: the identity operator leaves a quantity unchanged. (pages 62, 69, 74 & 183)	Operation Identity Operator addition $+\,0$ or $0\,+$ subtraction $-\,0$ multiplication $1(\ )$ or $(\ )1$ division $\dfrac{(\ )}{1}$
Increase Operator: an operator that represents addition. (page 38)	Adding 5 (or an increase of 5) is represented by the operator $+\,5$. The operator "$+\,5$" is read "an increase of five" or "plus five." It is <u>not</u> read as "positive five."
Integer: a natural number, zero, or the negative of a natural number. (page 245)	Examples of integers: 5, 0, -3 Examples of numbers that are not integers: $$\tfrac{1}{2},\ 1\tfrac{2}{3},\ -\tfrac{3}{4},\ \sqrt{5},\ 2\pi$$ The integers are also called whole numbers.
Invariant Principle for Differences: to keep the difference the same, if one term increases or decreases, the other term changes in the same way. (page 32)	$$x - y = (x + z) - (y + z)$$ $$x - y = (x - z) - (y - z)$$ <u>Idea</u> With two stacks of books, to keep the difference the same, if one goes down the other must go down the same amount. <u>Example</u> Solve for x: $73 - 25 = 63 - x$. *Solution*: 73 is decreased by 10 to get 63, so 25 must be decreased by 10 to get x. That is, $x = 25 - 10 = 15$.

Invariant Principle for Products: to keep the product the same, if one factor is multiplied by a number, the other factor must by divided by the same amount. (page 96)	$$xy = (xz)\left(\frac{y}{z}\right)$$ <u>Example</u> Solve for x: $5 \cdot 48 = 20 \cdot x$. *Solution*: 5 is multiplied by 4 to get 20, so 48 must be divided by 4 to get x. That is, $x = 48 \div 4 = 12$.
Invariant Principle for Quotients: to keep the quotient the same, if one part is multiplied or divided by a number, the other part must be changed in the same way. (pages 100 & 210)	$$\frac{x}{y} = \frac{x \cdot z}{y \cdot z} \quad \text{and} \quad \frac{x}{y} = \frac{\frac{x}{z}}{\frac{y}{z}}$$ <u>Example</u> Solve for x: $\frac{x}{12} = \frac{120}{6}$. *Solution*: 6 is multiplied by 2 to get 12, so 120 must be multiplied by 2 to get x. That is, $x = 120 \cdot 2 = 240$.
Invariant Principle for Sums: to keep the sum the same, if one addend increases or decreases, the other addend changes in the opposite way. (page 27)	$$x + y = (x + z) + (y - z)$$ <u>Idea</u> With two bags of marbles, to keep the sum of all the marbles the same, if one you take marbles out of one bag, you must add the same amount to the other bag. <u>Example</u> Solve for x: $68 + 47 = 65 + x$. *Solution*: 68 is decreased by 3 to get 65, so 47 must be increased by 3 to get x. That is, $x = 47 + 3 = 50$.
Inverse Operator: an operator that undoes the action of another operator. (pages 62 & 128)	<u>Operator</u> <u>Inverse Operator</u> $+3$ (an increase of 3) -3 (a decrease of 3) -5 (a decrease of 5) $+5$ (an increase of 5) $7(\ )$ (an expansion by 7) $\frac{(\ )}{7}$ (a contraction by 7) $\frac{(\ )}{6}$ (a contraction by 6) $6(\ )$ (an expansion by 6)

Joining an Operator: applying an operator to an operand (number or variable). (page 157)	When the operator 2( ) is joined to the operand x, the term $2x$ is obtained.
Least Common Multiple (LCM): the smallest of the *nonzero* common multiples of two or more numbers. (page 218)	Nonzero multiples of 6: 6, 12, 18, **24**, 30, 36, ... Nonzero multiples of 8: 8, 16, **24**, 32, 40, 48, ... The LCM of 6 and 8 is 24.
Lowest Terms: a fraction is in lowest terms when the only common factor of the numerator and the denominator is 1. (page 202)	When a fraction is in lowest terms, the Greatest Common Divisor (GCD) of the numerator and the denominator is 1. $\frac{3}{4}$ is in lowest terms since the GCD of 3 and 4 is 1. $\frac{6}{8}$ is not in lowest terms since the GCD of 6 and 8 is 2.
Magnitude: the size of an operator. (page 45)	Operator Magnitude $+ 3$ (an increase of 3) 3 $- 5$ (a decrease of 5) 5 2( ) (an expansion by 2) 2 $\frac{(\ )}{4}$ (a contraction by 4) 4
Mixed Number: the sum of an integer (whole number) and a proper fraction. (page 207)	Examples of mixed numbers: $3\frac{1}{2}$, $5\frac{7}{8}$ Note: $3\frac{1}{2}$ means $3 + \frac{1}{2}$.
Multiple: x is a multiple of y if x is evenly divisible by y. In other words, y must be a divisor (or factor) of x. (page 76)	The first 5 multiples of 3 are 0, 3, 6, 9, and 12: $0 \cdot 3 = 0$, $1 \cdot 3 = 3$, $2 \cdot 3 = 6$, $3 \cdot 3 = 9$, $4 \cdot 3 = 12$.
Multiplicative Identity: the multiplicative identity is 1 because multiplying by one makes no change. (page 69)	The multiplicative identity operator is 1( ).

Multiplicative Inverse: an operator that undoes the action of an expansion or a contraction operator. When referring to a number, it means the reciprocal. (pages 128)	The multiplicative inverse of $2(\ )$ is $\frac{(\ )}{2}$. When a multiplying by 2 is followed by a dividing by 2, the combined effect is no change.
Natural Numbers: the numbers $1, 2, 3, 4, 5, \ldots$ (page 1)	The natural numbers are the same as the positive integers. They are also called the positive whole numbers.
Numerator: the part of a fraction that is above the line. It is divided by the denominator (on the bottom). (page 2003)	In the fraction $\frac{3}{5}$, the numerator is 3.
Operand: the number, variable, or expression that is operated on by an operator. (page 38)	In the difference $5 - 2$, 5 is the operand and -2, a decrease of 2, is the operator. In the product $2(x + y)$, we may think of $x + y$ as the operand and $2(\ )$ as the operator. Or, we may think of 2 as the operand and $(\ )(x + y)$ as the operator.
Operator: an instruction to change a number. (page 38)	Operation — Sample Operator addition — $+2$ (an increase of 2) subtraction — -5 (a decrease of 5) multiplication — $4(\ )$ (an expansion by 4) division — $\frac{(\ )}{5}$ (a contraction by 5) exponentiation — $(\ )^3$ (raising to the third power)
Operator Method: a way of solving an equation by removing an operator from one side of an equation and joining the inverse operator to the other side. (pages 158 & 163)	Given the equation $x + 5 = 8$, remove the increase of 5 from the left side and join its inverse (a decrease of 5) to the right side: $$x + 5 = 8 \quad \rightarrow \quad x = 8 - 5$$ The effect is to subtract 5 from both sides.

Order of Operations: the rules that tell us what order to follow when performing arithmetic operations. (pages 19, 83 & 180)	When parentheses are used, do whatever is inside the parentheses first. Otherwise, start high and work down in the order of operations chart. Within a level, work left to right.

Exponentiation "power"	Taking a Root "root"
Multiplication "product"	Division "quotient"
Addition "sum"	Subtraction "difference"

Power: the name of the form when the last operation is exponentiation. (page 174)	$(5 - 3)^2$ is called a power because the last operation is exponentiation (raising to the second power). In the power 7^3, 7 is called the base and 3 is called the exponent.
Prime Factored Form: a way of writing a natural number greater than 1 as a product of its prime factors. It is unique when the primes increase from left to right and repeated primes are written as powers. (page 187)	The prime factored form for 90 is $2 \cdot 3^2 \cdot 5$. The primes (2, 3, and 5) increase from left to right, and the repeated factor of 3 is written as 3^2.
Prime Number: a natural number greater than 1 that has exactly two factors: itself and 1. (page 187)	7 is prime because its only factors are 1 and 7. 6 is not prime because 2 and 3 are factors of 6. If a natural number greater than 1 is not prime, it is called *composite*.
Product: the name of the form when the last operation is multiplication. (page 67)	The expression $(5 + 3)(2)$ is a product because you compute inside the parentheses first, then multiply.

Proper Fraction: a fraction in which the numerator is less than the denominator. (page 201)	These are proper fractions: $\frac{2}{3}$, $\frac{5}{8}$, and $\frac{3}{6}$. Note that $\frac{3}{6}$ is a proper fraction, even though it is not in lowest terms. These are not proper fractions: $\frac{5}{2}$, $\frac{3}{3}$, and $1\frac{3}{4}$.
Quotient: the name of the form when the last operation is division. (page 77)	The expression $\frac{5+3}{2}$ is a quotient. The division bar acts like parentheses: first compute $5 + 3$. Then divide.
Rational Number: a number that can be written as a fraction (a quotient of two integers). (pages 198 & 245)	Examples of rational numbers: 4, $\frac{2}{3}$, -17, 3.5 When a rational number is written as a decimal, the decimal either terminates or repeats.
Reciprocal: one of two numbers whose product is 1. (page 207)	To find the reciprocal of a fraction, interchange the numerator and the denominator. For example, the reciprocal of $\frac{3}{5}$ is $\frac{5}{3}$. The reciprocal of 3 is $\frac{1}{3}$.
Signed Numbers: the positive numbers, negative numbers, and zero. (page 244)	The positive signed numbers are to the right of zero on the number line and the negative signed numbers are to the left of zero.
Simple Fraction: a fraction $\frac{m}{n}$ where both m and n are counting numbers. (page 201)	$\frac{2}{3}$ and $\frac{7}{2}$ are simple fractions. $\frac{2\cdot 3}{4}$, $\frac{\frac{2}{3}}{4}$, and $\frac{5}{3+8}$ are not simple fractions.

Solve an equation (for a particular number or variable): find an equivalent equation in which the number (or variable) appears only once and it is by itself on one side of the equation. If an equation only has one variable, then solving the equation means finding the value of the variable that makes the equation true. (pages 28 & 136)	When the equation $5 + 4 = 9$ is solved for 4 without computing, we get $4 = 9 - 5$. When the equation $a + b = c$ is solved for b, we get $b = c - a$. When we solve the equation $x + 3 = 8$, we get $x = 5$.
Sum: the name of the form when the last operation is addition. (page 2)	The expression $5 + 2^3$ is a sum. The power is computed first, then the addition.
Symmetric Property: the two sides of an equation may be interchanged. (page 136)	If $x = y$, then $y = x$.
Three-Number Method: a way of solving an equation by studying the structure of the equation and seeing it as a sum and two addends or a product and two factors. (pages 138 & 148)	Here are the basic patterns: [sum] = [addend] + [addend] [addend] = [sum] – [addend] [product] = [factor] × [factor] [factor] = [product] ÷ [factor]
Transform: the final state or result when an operator is applied to an operand. (page 38)	When the operator 2( ) is applied to the operand $x + 5$, the transform is $2(x + 5)$.
Unique: the only one of its kind. (page 67)	In order for $\frac{x}{y}$ to be defined, there must be a unique (only one) number z such that $y \cdot z = x$.
Vector: a line segment that has both direction and length. (page 42)	Here is a vector model for an increase of 4:
Zero-factor Property: if a product of two numbers is zero, then one of the numbers must be zero. (page 68)	If $x \cdot y = 0$, then either $x = 0$ or $y = 0$.

Selected Answers

Chapter 1

Exercise 1.1
  1. True   **3.** True   **5.** $r + s$ **7.** $6 + 2$ **9.** 19   **11.** not possible   **13.** 53
15. 50   **17.** 47   **19.** 79

Exercise 1.2
  1. $6 + 6 = 12$ **3.** $0 + 0 = 0$ **5.** $14 + 14 = 28$ **7.** $15 + 15 = 30$

Exercise 1.3
  1. $4 + 5 = 9$ **3.** $5 + 2 = 7$ **5.** $0 + 5 = 5$ **7.** $5 + 4 = 9$ **9.** $6 = 6$ **11.** $2 + 15 = 17$

Exercise 1.4
  1. $6 = 2 + 4$ **3.** $7 = 1 + 6$ **5.** $12 = 3 + 9$ **7.** $18 = 14 + 4$ **9.** $25 = 21 + 4$
11. $23 = 23$

Exercise 1.5
  1. 5   **3.** 7   **5.** 8   **7.** $14 - 5 = 9$ **9.** $15 + 3 = 18$ **11.** $30 - 18 = 12$
13. $12 + 6 = n$ **15.** $x + m = 41$ **17.** $x + k = m$

Exercise 1.6
  1. 6   **3.** 3   **5.** not defined   **7.** 3   **9.** not defined   **11.** 7

Exercise 1.7
  1. $(7 + b) + a,\ a + (b + 7),\ (b + 7) + a$ **3.** $(n - 5) + m$ **5.** $6 + (x - y)$
  7. (a) to (b) commutative, (b) to (c) associative, (c) to (d) commutative, (d) to (e) commutative, (e) to (f) associative

Exercise 1.8
  1. $135 + 100$ **3.** $100 + 253$ **5.** $x = 548 - 2 = 546$ **7.** $x = 567 + 4 = 571$

Exercise 1.9

1. $145 - 100$ **3.** $100 + 172$ **5.** $165 - 50$ **7.** $x = 434 - 2 = 432$ **9.** $x = 572 + 10 = 582$

11. (a) $413 + 349 = 425 + x$ (b) $x = 349 - 12 = 337$

Chapter 2

Exercise 2.1

1. $x, +9$ or $9, x+$ **3.** $x, -3$ **5.** $16, -x$ **7.** $b + 7, -x$ **9.** $x, -(3 + c)$

11. $w + 5, +m$ or $m, (w + 5) +$

Exercise 2.2

1. (a) $+6$ (b) $+4$ (c) $+10$ (d)

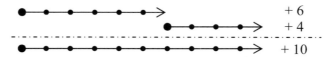

3. (a) -5 (b) -4 (c) -9 (d)

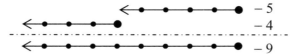

5.

7.

9.

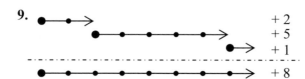

11. $+9$ **13.** -13 **15.** $+14$

Exercise 2.3

1. $18 - (3 + 4)$ or $18 - (4 + 3)$ **3.** $17 + (2 + 8)$ or $17 + (8 + 2)$

5. $25 - (m + n)$ or $25 - (n + m)$ **7.** $x - (a + 5)$ or $x - (5 + a)$ **9.** $x + (n + 3)$ or $x + (3 + n)$

Exercise 2.4

1. (a) $+6$ (b) -4 (c) $+2$ (d)

3. (a) − 8   (b) + 4   (c) − 4   (d)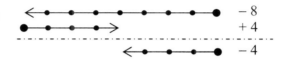

5. + 2 / + 3 / + 5

7. − 2 / + 7 / + 5

9. − 3 / + 2 / − 1

11. − 3   **13.** + 17   **15.** + 12   **17.** − 14

19. + 7 + 9 − 5 = + 11

Exercise 2.5

1. $15 − (3 + 7)$ or $15 − (7 + 3)$ **3.** $15 − (7 − 3)$ **5.** $15 + (7 − 3)$

7. $15 + (3 + 7)$ or $15 + (7 + 3)$ **9.** $30 − (m + n)$ or $30 − (n + m)$ **11.** $30 + (m − n)$

13. $30 − (m − n)$

Exercise 2.6

1. − 8   **3.** + 7   **5.** − 6   **7.** + 7

Chapter 3

Exercise 3.1

1. $4 \cdot 9$ **3.** 40   **5.** 24   **7.** 32   **9.** 45   **11.** 24   **13.** not possible

15. $0 + 6 + 6$ **17.** $4(\ )$ **19.** $6(\ )$

Exercise 3.2

1. $7; \dfrac{(\ )}{n}$ **3.** $4; (\ )m$ or $m; 4(\ )$ **5.** $8; (x − 3)(\ )$ or $x − 3; (\ )8$ **7.** $4 \cdot 17 = 68$

9. $\dfrac{322}{23} = 14$ **11.** $6 \cdot a = c$ **13.** $b \cdot (c + d) = a$ **15.** $\dfrac{(\ )}{2}$

17. undefined

Exercise 3.3

1. $0, 1, 2, 3$ **3.** $0, 2, 4, 6$ **5.** $0, 4, 8, 12$ **7.** $1, 2, 5, 10$ **9.** $0, 10, 20, 30$

11. 0, 15, 30, 45   **13.** 2, 4, 5, 10, 20

Exercise 3.4

1. 2   **3.** 3, 5   **5.** none of these   **7.** 1, 3, 7, 21   **9.** 1, 2, 4, 5, 8, 10, 20, 40

Exercise 3.5

1. difference, 9   **3.** sum, 22   **5.** quotient, 12   **7.** sum, 27   **9.** product, 25

Exercise 3.6

1. $(10 - 3)8$ **3.** none   **5.** $4 + \dfrac{x}{5}$ **7.** $\left(\dfrac{a}{b}\right)5$ **9.** $\dfrac{a + x}{3}$ **11.** $(y)(3) - 5$

13. (a) to (b) commutative, (b) to (c) associative, (c) to (d) commutative, (d) to (e) associative, (e) to (f) associative

Exercise 3.7

1. $(6)(7) + (6)(2)$ **3.** $\dfrac{27}{3} - \dfrac{12}{3}$ **5.** $\dfrac{y}{7} + \dfrac{x}{7}$ **7.** $(a)(5) + (4)(5)$ **9.** $9(7 - 3)$

11. $(8 - 3)y$ **13.** $\dfrac{x - h}{3}$ **15.** $6(a - b)$

Exercise 3.8

1. $x = 14 \div 2 = 7$ **3.** $x = 16 \cdot 2 = 32$ **5.** $x = 8 \cdot 4 = 32$ **7.** $x = 15 \cdot 3 = 45$

Exercise 3.9

1. $\dfrac{48}{16}$ **3.** $x = 35 \cdot 2 = 70$ **5.** $x = 2 \cdot 5 = 10$ **7.** $x = 22 \div 2 = 11$ **9.** $x = 14 \cdot 3 = 42$

11. $x = 24 \div 4 = 6$

Chapter 4
Exercise 4.1

1. Expand by 2 and then expand by 4.   **3.** 24( )   **5.** $\dfrac{(\ )}{16}$ **7.** $\dfrac{(\ )}{4}$

9. $mn(\ )$

Exercise 4.2

1.
$$4\left(\dfrac{(\ )}{2}\right)\left(\begin{matrix}\dfrac{(\ )}{2}\\ 4(\ )\end{matrix}\right.$$

3. Expand by 12 and then contract by 4. 3( )

5. Expand by 3 and then contract by 8. Not defined

7. Contract by 12 and then contract by 4. $\dfrac{(\ )}{48}$

9. Contract by 3 and then expand by 15. 5( )   **11.** Contract by 3 and then expand by 5. Not defined

13. Contract by 2 and then expand by 8. 4( )

Exercise 4.3

1. Contract by 15 and then expand by 3.   **3.** Contract by 4 and then expand by 20.

5. Expand by 6 and then contract by 18.   **7.** $\dfrac{3(\ )}{15}$ **9.** $\dfrac{20(\ )}{4}$ **11.** $6\left(\dfrac{(\ )}{18}\right)$ **13.** $\dfrac{(\ )}{5}$

15. 5( )   **17.** $\dfrac{(\ )}{3}$

Exercise 4.4

1. $\dfrac{17+1}{3}-2$, difference   **3.** $\dfrac{3(8+4)}{2}$, quotient   **5.** $\dfrac{12-2}{5}+6$, sum

7. $\left(\dfrac{18}{2}-4\right)3$, product   **9.** $\dfrac{15-3}{4}+7$, sum

Exercise 4.5

1. $12;\ -5$ **3.** $9;\ -3,\ 4(\ )$ **5.** $5;\ (\ )6,\ -7$ or $6;\ 5(\ ),\ -7$ **7.** $10;\ \dfrac{(\ )}{2},\ 6+$

9. $a;\ \dfrac{(\ )}{b},\ +c$ **11.** $f;\ +g,\ \dfrac{(\ )}{h}$ or $g;\ f+,\ \dfrac{(\ )}{h}$

Exercise 4.6

1. $+12$ **3.** -9 **5.** x **7.** 6   **9.** not possible   **11.** not possible   **13.** $x+3$

Chapter 5

Exercise 5.1

1. (S): 12; (A): 9, 3   **3.** (S): 17; (A): 6, 11   **5.** (S): 16; (A): 4, 12

7. $9 = 12 - 3, \ 3 = 12 - 9$ **9.** $17 = 6 + 11, \ 6 = 17 - 11$ **11.** $4 = 16 - 12, \ 12 = 16 - 4$

Exercise 5.2

1. 41   **3.** x **5.** $\dfrac{a}{b}$ **7.** $x = 41 - a$ **9.** $x = 5m + 32$ **11.** $x = \dfrac{a}{b} - 6$ **13.** $x = 7$

15. $x = 19$ **17.** $x = 5$

Exercise 5.3

1. (P): 24; (F): 3, 8   **3.** (P): 18; (F): 2, 9   **5.** (P): 96; (F): 8, 12   **7.** $3 = \dfrac{24}{8}, \ 8 = \dfrac{24}{3}$

9. $2 = \dfrac{18}{9}, \ 18 = 2 \cdot 9$ **11.** $8 = \dfrac{96}{12}, \ 12 = \dfrac{96}{8}$

Exercise 5.4

1. (P): n; (F): 5, y **3.** (P): y; (F): 8, $a + 3$ **5.** (P): $n + 6$; (F): y, m **7.** $y = \dfrac{n}{5}$

9. $y = 8(a + 3)$ **11.** $y = \dfrac{n + 6}{m}$ **13.** $x = 3$ **15.** $x = 28$ **17.** $x = 4$

Exercise 5.5

1. $x = 12$; $12 + 3 = 15$ **3.** $x = 28$; $18 = 28 - 10$ **5.** $x = 5$; $9 = 4 + 5$ **7.** $x = n - 4$

9. $x = 6 - 2y$ **11.** $x = cd + 3y$

Exercise 5.6

1. $x = 4$ **3.** $x = \dfrac{m}{5}$ **5.** $x = 5y$ **7.** $x = \dfrac{t}{m + 3}$ **9.** $x = \dfrac{n}{3} + 5$ **11.** $x = \dfrac{n + 4}{7}$

Exercise 5.7

1. $x = k + 3$ **3.** $x = \dfrac{n}{7}$ **5.** $x = 18 - a$ **7.** (S): $\dfrac{a}{5}$ and (P): a **9.** (S): n and (P): x

11. (S): y and (S): $y - x$ **13.** $x = \dfrac{a}{5} - 3$ **15.** $x = 4(n - 7)$ **17.** $x = y - (a + 5)$

Chapter 6

Exercise 6.1

1. $1 \cdot 2 \cdot 2 \cdot 2 \cdot 2 = 16$ **3.** $1 \cdot 5 \cdot 5 = 25$ **5.** 36   **7.** 1   **9.** 81   **11.** 64   **13.** 5^4
15. d^3

Exercise 6.2

1. sum, 25   **3.** power, 9   **5.** product, 18   **7.** sum, 27   **9.** difference, 4   **11.** 8

Exercise 6.3

1. 5^6 **3.** 8^4 **5.** 4^6 **7.** 5^3

Exercise 6.4

1. $3 \cdot 7$ **3.** 23   **5.** 5^2 **7.** 3^3 **9.** 29   **11.** 2^5

Exercise 6.5

1. GCF $= 3$ and LCM $= 2 \cdot 3^2 \cdot 5^2$ **3.** GCF $= 2 \cdot 7$ and LCM $= 2^2 \cdot 3^2 \cdot 7$

5. GCF $= 5$ and LCM $= 2^2 \cdot 3 \cdot 5^2$ **7.** GCF $= 2 \cdot 3^2$ and LCM $= 2^2 \cdot 3^4 \cdot 5 \cdot 7$

9. 360 seconds or 6 minutes

Chapter 7

Exercise 7.1

1. Expand by 4 and contract by 5; smaller   **3.** Expand by 5 and contract by 3; larger

5. Expand by 2 and contract by 9; smaller   **7.** $\dfrac{3}{6}$ and $\dfrac{7}{8}$ **9.** $\dfrac{3}{4}$ **11.** $\dfrac{5}{7}$ **13.** 7

15. $\dfrac{1}{2}$

Exercise 7.2

1. $\dfrac{1}{4}$ **3.** $\dfrac{8}{15}$ **5.** $\dfrac{2}{3}$ **7.** $\dfrac{1}{9}$ **9.** $\dfrac{4}{9}$ **11.** $6\dfrac{2}{3}$ **13.** $\dfrac{8}{3}$ **15.** $\dfrac{1}{8}$

Exercise 7.3

1. $\dfrac{7}{10}$ **3.** $\dfrac{4}{5}$ **5.** $\dfrac{4}{5}$ **7.** $\dfrac{5}{21}$ **9.** $x = \dfrac{1}{4}$ **11.** $x = \dfrac{6}{5}$ **13.** $x = \dfrac{4}{5}$

Exercise 7.4

1. $1\dfrac{4}{15}$ **3.** $\dfrac{1}{10}$ **5.** $\dfrac{17}{30}$ **7.** $\dfrac{11}{30}$ **9.** $3\dfrac{5}{6}$ **11.** $x = \dfrac{7}{12}$ **13.** $x = \dfrac{11}{15}$

15. $x = \dfrac{17}{30}$

Exercise 7.5

1. $\dfrac{4}{9}$ **3.** $\dfrac{2}{5}$ **5.** $\dfrac{5}{9}$ **7.** $\dfrac{4}{5}$ **9.** $\dfrac{13}{60}, \dfrac{14}{60}$ **11.** $\dfrac{43}{63}, \dfrac{44}{63}$

Chapter 8

Exercise 8.1

1. $\dfrac{29}{100}$ **3.** $\dfrac{23}{500}$ **5.** $\dfrac{13}{5}$ **7.** 0.37   **9.** 0.6   **11.** 0.36   **13.** 1.72   **15.** 2.7

Exercise 8.2

1. 35.77   **3.** 12.23   **5.** 8.72   **7.** 9.42   **9.** 96   **11.** 3.9

Exercise 8.3

1. 48,540   **3.** 49,000   **5.** 126,080   **7.** 126,000   **9.** 6   **11.** 10   **13.** 800

15. 11

Exercise 8.4

1. 5   **3.** −12   **5.** $-\dfrac{7}{12}$ **7.** −5.6   **9.** −2.3

Exercise 8.5

1. $\dfrac{11}{3}$, −2, 3.9   **3.** $x < 2$ **5.** $n \geq 7$

7. $x < 4$ **9.** $n > 8$

Index